Centralization and Dispersion of Space
空间的聚散
Wang Yun
王昀 著

中国建筑工业出版社

目录		Contents
前言	1	Preface
导言	2	Introduction
"西溪学社"风景	8	Scene of "Xixi Institute"
"西溪学社"离散式会所的整体布局	32	Overall layout of "Xixi Institute", a dispersed club
构成"西溪学社"离散式会所的建筑	41	Architecture consist of "Xixi Institute"
1."方体空间"	42	"Cube Space"
2."梅花"	52	"Plum Flower"
3."喇叭"	70	"Trumpet"
4."空中立方体"	86	"Overhead Cube"
5."烟囱"	92	"Chimney"
6."椭圆住宅"	108	"Oval Residence"
7."方体咖啡"	120	"Cube Café"
8."桥宅"	124	"Bridge Residence"
9."长宅"	136	"Long Residence"
10."长方体"	152	"Cuboid"
11."水上漂"	162	"Water Float"
12."框景舞台"	182	"Enframed Stage"
13."扇形剧场"	188	"Fan-shaped Theatre"

前言

坐落在杭州西溪湿地的这组离散式布局建筑群,是一个以"西溪学社"为名的文化交流场所。该设计开始于2007年10月,2009年12月基本完成。由于湿地本身环境的特殊性,设计时将彼此使用目的不同的10个建筑和6个构筑物散落布置在分散状的地块环境中,进而客观地构成了由若干零散附属建筑散落在建设基地上的离散式的聚落形态。基地中的建筑由"六和塔"、"水上漂"、"空中立方体"、"梅花"、"喇叭"以及"方体空间"等单体建筑所构成,自身在满足使用目的的同时,力图使整个聚落成为一个能够激发使用者风景观念的装置。

从某种意义上说"离散"是一种空间聚合状态的显现,并与当下的社会状态相应和。网络技术的普及使人的独立状态获得极大的体现,借助于电子网络,物理距离已经不再成为人与人之间真正的距离状态。西溪湿地这组建筑群中建筑与建筑之间的距离状态的表现或许与当下社会状态直接应和。

有意味的是,整个"西溪学社"的建造过程以及当下这组建筑所处的状态,事实上同样经历了"聚合"和"离散"的过程。集群设计的初始聚合,过程中部分建筑师作品没能开建,以及最终自己也不得不从心理上离散这组建筑的结果,与该基地的离散性提示如出一辙。

当这本书得以与诸位见面的时候,现场或已成为废墟,或已发生了面目全非的改变。而今天所有由这组建筑所呈现的物理性的离散状态,对我而言或许能从观念上获得一个"为了忘却"的意义。

Preface

The design for this group of dispersed buildings situated in Hangzhou Xixi Wetland started in October 2007 and its construction basically completed in December 2009. The total building area I was involved covered about 3800 square meters. As to the main purpose of use, the owner hoped to establish a venue for cultural exchange, with the name of "Xixi Institute". Due to its own special wetland environment, the land lots were dispersed instead of concentrated. In architectural design, we took advantage of this special environment and scattered 10 buildings and 6 structures of different usage across the land lots. Therefore, a dispersed community pattern was objectively formed, with several affiliated structures scattered across the base. Architecture on the base consists of single buildings such as "Six Harmonies Pagoda", "Water Float", "Overhead Cube", "Plum Flower"," Trumpet", "Cube Space" and so on, trying to make this whole community into a facility that can inspire users to appreciate scenery.

Being "dispersed" is not only a kind of spatial community pattern but also actually a contemporary social status. The development of science and technology, particularly the popularity of network technology, enables maximum demonstration of a person's individual independence. By means of electronic network, physical distance is no more the same as actual distance between people. Despite the dispersed distance between buildings of this architectural group in Xixi Wetland, they form an integral whole. Observed from the same perspective, the overall architectural work, designed by 12 architects who participated in designing this area, also form a dispersed relation among each other on a larger scale.

This process of concentration and dispersion is also reflected in designing and building this art village as well as in the current status of this architectural group. From the initial stage of group design concentration, to the stage when some architects did not start their construction, to the final result that buildings had to dispersed, the whole situation is very similar to the condition of this base at the very beginning.

The publishing of this book today actually is trying to showcase the best condition of this project. Even if possibly the current site is already in ruins or undergoes fundamental change, this for me maybe a "memorial meant to forget".

导言
Introduction

一、关于项目

"西溪学社"是杭州西溪湿地第三期艺术村中的一个局部。使用者希望在这里能够为文化和艺术界人士提供一个进行艺术创作和学术交流的场所。在整个艺术村设计的初始,杭州"一创意"策划公司的黄石先生召集了国内12位青年建筑师共同参与了这个项目。每一位建筑师利用不同的地块,分别设计不同使用目的的建筑,有美术馆、度假酒店、会所等。

我所负责的区域,业主希望在这里建一组名为"西溪学社"的文化交流和艺术创作场所。而引发这一想法的缘由据说是受杭州西湖北侧西泠印社的启发,西泠印社曾在中国近代文化发展史上有过很大的影响。基于这样的思考,业主希望在西溪湿地中同样建一个能传承中国当代文化和艺术的场所,并希望学社建成后可不定期地请一些艺术家和学者来这里进行创作并为学生授课,同时这组建筑也为艺术家本人提供一个怡人的创作环境。西溪学社的建设基地很特殊,用地分散且彼此间被既有的水塘所分隔。业主希望整体的建筑面积能够控制在3800m²左右,同时能够满足学社的使用需求。

"西溪学社"整体的建筑是由10个建筑和6个构筑物所组成,其总用地范围为5.98万m²,其中作为具体使用的学社总建筑面积为3835m²。

I. About the project

"Xixi Institute" is part of the Art Village amid the third phase of Hangzhou Xixi Wetland. Users hope to provide people in the circle of culture and art with a venue right here to conduct artistic creation and academic exchange. At the initial stage of design for the whole Art Village, Mr. Huang Shi of Hangzhou "One Innovation" Design Company invited 12 young architects in China to jointly participate in this project. Each architect utilized a different land lot to separately design building for different purposes, such as art gallery, resort hotel, club and so on.

As to the land lot I was in charge, the owner hoped to build a venue named as "Xixi Institute" for cultural exchange and artistic creation. This thought was said to be inspired by Xiling Seal Engravers Society located on the north side of the West Lake in Hangzhou, which once had great influence on modern culture development history in China. Based on such thought, the owner hoped to build a similar venue capable of inheriting Chinese modern culture and art amid Xixi Wetland, which may not only attract some artists and scholars to conduct creative work and teach from time to time but also provide pleasant environment conducive to creative work for artists themselves.

The construction base of "Xixi Institute" was quite unique, as the land pieces were scattered and separated by existing ponds. The owner expected that while the project can satisfy the usage need of the institute, the total built-up area can be controlled within 3800 square meters.

图1 中国云南丙中洛"雾里村"的离散式布局
Figure 1: Dispersed layout of "Wuli Village" in Bingzhongluo, Yunnan Province, China

层的占地面积为2710m², 除个别建筑3层外, 多数由2层的建筑物所构成。

二、设计中的思考
1. 离散式布局

当第一次看到这个地块时, 感觉建筑用地特别分散。如何满足既有集中又有分散的功能需求, 满足创作、教学、居住和展览的需要是设计时应重点考虑的。在西溪湿地这样一个怡人的自然景观中, 集中地盖几千平方米的建筑, 体量将是非常大的, 如何处理好建筑与环境的关系是设计的一个难点。考虑到传统聚落形态中, 尽管大量的聚落布局采用的是集中式的聚合状态, 但为数不多的"离散式"布局状态却能巧妙地处理好体量与基地的关系。

所谓离散式是指彼此分离, 但又相互关联的一种布局状态。类似于游牧性质的空间体系, 个体之间的距离比较分散, 但其中又并不缺乏逻辑性, 如图1所示的中国"雾里村"（又名"五里村"）, 便是一种明显的"离散式"聚落布局。聚落中户与户之间彼此距离较大, 而整体共同地支配一个较大范围的区域。

2. "经验"与"选好"

实际上, 设计是一个非常个人化的活动过程, 在我们看来, 设计过程整体的完成与设

The overall project of "Xixi Institute" consists of ten buildings and six structures, with a total land area of 59800 square meters containing specific institute area of 3835 square meters. The first floor occupies a land area of 2710 square meters. Most architects have two storeys except a few three-storeyed architects.

II. Designing thoughts
1. Dispersed layout

The designer considers the land lot much dispersed at the first glance. The key points to be considered in designing are how to satisfy the functional need of both centralization and dispersion, how to meet the need of creation, teaching, living and exhibition. Amid the pleasant natural scene of Xixi Wetland, an architectural project with land area of several thousand square meters will be quite massy, therefore it is difficult to handle the relationship between the project and surroundings. Considering in the traditional community pattern, we find that although most communities adopt centralized pattern, a few "dispersed" patterns can tactfully handle the relationship between mass and base.

Dispersed pattern means a status in which things are not only separated from each other but also interrelated to each other. It is similar to a nomadic spatial system, in which individuals are scattered in terms of distance but bonded by certain logic relationship. As shown by Figure 1, the community layout of "Wuli Village" in China demonstrates a typical "dispersed" pattern. Each household is far from the other while jointly supporting a community within large scope.

图2 中国青海的"日月山"聚落
Figure 2: "Riyueshan" Community in Qinghai Province, China

图3 中国青海"日月山"聚落的平面图
Figure 3: Plan of "Riyueshan" Community in Qinghai Province, China

计者的个人经验密切相关,设计者的这些"经验",在设计时会依照"选好"的倾向而流淌出来。鉴于这样的思考,在进行设计时,设计师所设计的结果,与设计师个人的经历发生着密切的关联。

3. "引用"与"转化"

西溪学社的南侧布置有一组小旅馆性质的建筑,这是将我曾经去调查和测绘过的名为"日月山"的村聚落(图2)进行直接引用而完成的设计。设计时试图将"日月山"聚落的平面图(图3)中的空间组成直接转化成新设计的小宾馆的空间关系。在新的设计中采用了现代的材料,而非将原有的夯土墙元素照搬过来。原有的20m×20m左右的住居单元,在新旅馆的设计中转化为4m×4m的一组小方单元的客房建筑,整个小宾馆一共有24间客房。可以容纳一个班的学生,且每个人都可以拥有一个独立的住居客房。

4. 经验与意向

在"西溪学社"这个区域中,我们还设置了几个可以满足师傅带徒弟的小型工作室。在这些工作室中,可以进行不同场景的活动,或是在梅花形的曲墙空间内聚谈,体验安静的

2. "Experience" and "preference"
Actually, designing is a procedure with much personalized activities. In our view, the completion of the overall designing procedure is closely related to a designer's individual "experience", which will flow out during designing according to "preference". Based on such consideration, during designing a designer's result is closely correlated with his or her personal experience.

3. "Reference" and "transformation"
On the south side of Xixi Institute there is a group of buildings with the feature of inn, which is my design with direct reference of "Riyueshan" village community that I visited and surveyed before (Figure 2). During design I try to directly transform space in the plan of "Riyueshan" community (Figure 3) into the spatial relationship of newly designed inns. In the new design, modern materials instead of the original loam wall are applied. The original living unit of 20m×20m is transformed into small square unit of 4m×4m during designing the new inn. There are totally 24 rooms in the inn to accommodate a class of students, each enjoying an independent room.

4. Experience and Intention
Within the area of "Xixi Institute", we also establish several small studios where masters can teach their students. In these studios people can conduct activities of various scenarios, such as chatting within the plum-flower shape walls to enjoy serene atmosphere, or holding liberal concerts within outdoor amphitheatre. Different venues set possibilities for different stories to develop.

图4 "西溪学社"的"大烟囱"建筑
Figure 4: "Big Chimney" building of "Xixi Institute"

图5 杭州"六和塔"旧影
Figure 5: Old photo of "Six Harmonies Pagoda" in Hangzhou

气氛；或是在露天圆形剧场内举办自由的音乐会。不同场所，设定了不同故事展开的可能性。

　　大烟囱下的建筑，是一个可以举办展览的场所。大的台阶与江浙一带常见的桥相关联。在设计的结果中可发现里面有很多细节实际上与作为设计者的"我"所曾游历过的杭州场景印象及记忆相关联。"大烟囱"（图4）或许可以唤起"六和塔"的意象（图5）。通过建筑让人体验一种似曾相识的感觉，唤起记忆中沉睡的景象。不仅仅是形式上的联想，而是更多地把头脑中的场景重新建构和搭接起来。而这或许是建筑空间中所具有的一个拥有重要意义的特征。

　　建筑本身并不单纯地存在于三维的"透视学"意义的层面，而更是存在于时间和记忆的层面，存在于游走的过程中。由散步者脑海中一个个叠加起来的场景而最终形成的印象和感受，在我看来才是建筑最本质的东西。从这个层面上讲，建筑本身可能更接近于电影，因为对电影而言你不可能通过一个剧照来判断整个电影的故事，而必须把电影全部看完才能获得完整的感受。

　　在江南的水乡聚落中行走，架在水上的"桥"带给人重要的体验（图6），先上桥再下

The building below the big Chimney is available for exhibitions. Big staircase is correlated with bridges common in Jiangsu and Zhejiang. Many details found in the result of design are actually correlated with my impression and memory about Hangzhou scenes that I as the designer once experienced."The Big Chimney" (Figure 4) may evoke impression of the "Six Harmonies Pagoda" (Figure 5). People can experience a kind of déjà vu feeling through building and arouse dormant scenes in their memory. It is not only a mental association in form but also a reconstruction and bridging of scenes in memory, which may be a significant feature belonging to architectural space.

Architecture itself does not simply exist on the level of three-dimensional "perspective science" but rather on the level of time and memory. The essential factors of architecture, in my opinion, are impression and feeling formed finally by superimposed scenes in a walker's mind during the procedure of such walks. On this level, architecture itself may be more similar to a movie, as you must finish the whole movie to get full impression instead of judging the whole story of the movie by a single stage photo.

When one walks in waterside communities of Jiangnan, "bridges" over water offers important experience (Figure 6). I consider it a cultural representation truly transcending regions when one walks up and then down a bridge, with experience and memory from feet and body. It is a significant thinking and method for us to continue history by modern factors, offering linkage between modern

图6 杭州的西湖边上的"桥"
Figure 6: "Bridge" by the West Lake in Hangzhou

图7 西溪湿地"桥宅"
Figure 7: "Bridge Residence" in Xixi Wetland

桥,我认为这种来自脚下和身体上的经验记忆是真正从形式上超越了地域的文化表征。将现代和历史不仅在视觉上,更通过体验产生内在的联想与碰撞,从而获得链接,这是我们用现代的要素使历史延续的一种重要思想和手段,也是超越从视觉上模仿传统的后现代符号式表述历史的观念表达(图7)。

三、超越狭隘地域性表达的可能性
1. 具象与抽象

在设计时,如何让人看上去有当地的地方性,而又并非单纯利用某一个地方性的形式去表达,是"西溪学社"这组建筑在设计中所进行思考的重点。如前述,这种唤起记忆中场景的做法,是解决这一困惑的基本出发点,为此,在具体的设计中,采用不确定的具体形式。换言之,采用抽象的要素作为基本的设计表达语言是设计时的重要选择,选择这一做法是建立在对于数学的抽象性理解的基础上完成的。比如"3只铅笔"和"3头牛"之间是没有可比性的,同时两个概念本身所传递的信息又是具体的、单一意义的,因为"铅笔"和"牛"之间是一个具象的存在,而当我们抛弃了"铅笔"和"牛"这种具象概念的存在,"3只铅笔"和"3头牛"之间就被赋予了共同的

and history after internal mental association and collision through experience instead of visual sense. It is also an expression of ideas to present history in post-modern symbolic ways, transcending the visual way of imitating tradition (Figure 7).

III. Possibility of expression transcending narrow regional features
1. Concrete and abstract

It is a key point in designing "Xixi Institute" buildings to give them regional features but not just by solely one certain local form. As mentioned earlier, we try to evoke scenes in memory, which is a basic starting point to solve this puzzle. Therefore, in concrete design, we adopt unspecific concrete form. In other words, it is an important choice during designing to adopt abstract factors as basic language in design expression. Choosing this method is based on understanding of abstractness in mathematics. For example, there is no comparison between "3 pencils" and "3 cows", meanwhile information conveyed by these two concepts themselves is concrete and singular, as both "pencils" and "cows" are concrete existence. But when we discard concrete concept of "pencil" and "cow", then we give common feature to "3 pencils" and "3 cows", with general and universal meaning of "3". The existence of universal abstractness enables rich mental association from a single abstract object.

2. Abstractness and multi-meaning interpretation

Figure 8 shows the scene of "Three Pools Mirroring the Moon" in Hangzhou, which is a kind of typical spatial scene in local regional culture. The

图8 杭州西湖的"水榭"
Figure 8: "Waterside Pavilion" of the West Lake

图9 "西溪学社"的"水上漂"建筑
Figure 9: "Water Float" building of "Xixi Institute"

特征,并具有广泛和普遍意义的特征,即都是"3"。这种具有普世价值的抽象性的存在,可以在一个单纯的抽象的对象物中获得丰富的联想。

2. 抽象性与多意性读解

图8是杭州的"三潭印月"的风景照。这种风景成为当地地域文化中一种具有代表性的空间场景。我们设计中的"水上漂"建筑是临水或是直接在水中建造起来的(图9)。当时造这栋房子的时候,实际上是对杭州"三潭印月"的意向表达,漂浮在水面上令人感受到神话般的意境。希望通过"水上漂"建筑而获得"水榭"的联想。"水上漂"后面作为衬景的树木,应该说源于对西湖的整体印象,尽管语言表达系统不同,我想有过西湖游览经验的人可以或能够感受到"三潭印月"的景象。不过,对我而言,更加关心的是没有见过"三潭印月"的人能够在这个场景中唤起他所能够看到的"那些"或"哪些"场景,而当这种歧义场景被"唤起"的瞬间,或许,不!那便一定是超越狭隘地域性的开始。

building of "Water Float" (Figure 9) that we design is bordering water or directly above water. When this building was being constructed, it served as expression of "Three Pools Mirroring the Moon" in Hangzhou, which floats above water with fairytale flavor. This building is expected to evoke mental association of "waterside pavilion". Trees behind this building, serving as background, originate from the overall impression about the West Lake. In spite of difference in lingual expression system, I suppose that people with experience of the West Lake can or may feel the scene of "Three Pools Mirroring the Moon". However, for me I care more about "those scenes" or "what scenes" can be aroused by this view particularly in minds of people who have never seen "Three Pools Mirroring the Moon". Once those different scenes are aroused, it maybe, oh no, then it must be the start of transcending narrow regional features.

"西溪学社"风景
Scene of "Xixi Institute"

从基地的西侧看"西溪学社"离散式会所
View of "Xixi Institute", a dispersed style club, from the west side of the base

从基地的北侧看"六和塔"意味的"大烟囱"
View of "Big Chimney", embodying implication of "Six Harmonies Pagoda", from the north side of the base

从基地的西侧看"西溪学社"离散式会所
View of "Xixi Institute", a dispersed style club, from the west side of the base

漂浮在水面上的"西溪学社"建筑群
Group of buildings of "Xixi Institute" floating above water

"水上漂"风景
Scenes of "Water Float"

"梅花"会所的南侧风景
Scenes on the south side of "Plum Flower" club

"西溪学社"东南角风景
Scenes at the south-east corner of "Xixi Institute"

"西溪学社" 东北角风景
Scenes at the north-east corner of "Xixi Institute"

"长宅"风景
Scenes of "Long Residence"

"西溪学社"基地内风景
Scenes of "Xixi Institute"

离散在西溪湿地中的"西溪学社"建筑群
Group of buildings of "Xixi Institute" scattered around Xixi Wetland

隐现于湿地中的"西溪学社"
"Xixi Institute" appearing indistinctly in the wetland

"西溪学社"离散式会所的整体布局
Overall layout of "Xixi Institute", a dispersed club

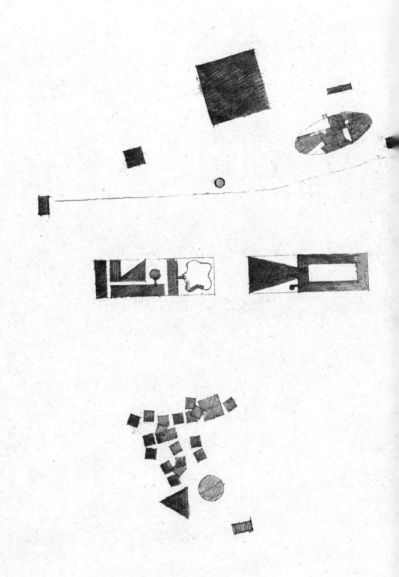

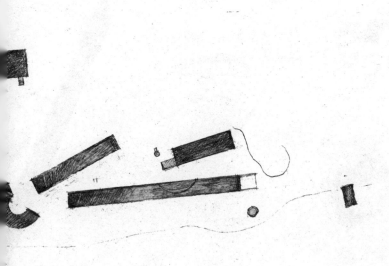

"西溪学社"整体构思草图
Sketch of "Xixi Institute"

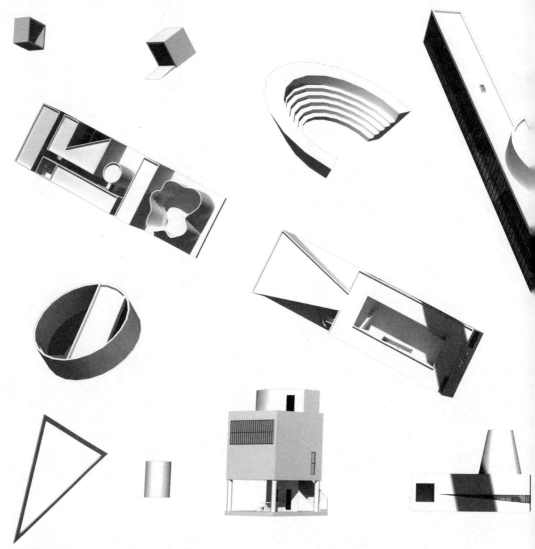

"西溪学社"整体的建筑设计采用明确的几何体组合，使其拥有共同幻想的特征。设计时在采用简洁几何体进行布局的同时，注重了聚落中所必需的"微差"设置，目的是使体验者能够在若即若离中体会到一种丰富性的存在。鉴于基地中可用于建设的场地较为分散，因此对建筑体块进行切分是进行该项目整体布局时所采用的手段。将大而整的体量打散，形成离散式的空间布局，进而也使建筑与自然之间得以更紧密地融合。离散是彼此分开，但离散式布局彼此之间

The overall architectural design of "Xixi Institute" adopts explicit geometry combination to possess features of common fantasy. The designers, while adopting simple geometry in layout, also pay great attention to "minor difference" setup necessary in a community so as to give a feeling of richness in a semi-close and semi-distant status. As land lots in the base suitable for construction are quite scattered, the designers cut and allocate architectural blocks in overall planning. By breaking huge and integral blocks, dispersed spatial layout is formed and the relationship between architecture and nature becomes more closely integrated. Being dispersed is being separated, but each factor in a dispersed layout

具有连带性关系。这种类似于游牧性质的空间体系，实际上与当代的状态有明显的契合关系。"西溪学社"用地分散，客观上提供了将建筑的体量得以缩小的前提。考虑到湿地中树木的尺度，将建筑体量拆分，能保证湿地本身原有的尺度体系。从这个意义上讲，采用传统聚落中所展示出的智慧是解决问题的有效手段。在设计时我们以"共同幻想"作为整体塑造的基础，依据环境特征强调微差的存在。进而使"西溪学社"这组离散的聚落成为一个有机的整体。

has connectivity. The spatial system similar to nomadic feature actually corresponds to the current status in an obvious way. Land lots of "Xixi Institute" are scattered, which serves as precondition to reduce architectural blocks. Considering the size of trees in wetland, we break up architectural mass to maintain the original scale system of wetland itself. From this perspective, it is an effective method to adopt wisdom reflected in traditional communities to solve problems. In design, we use "common fantasy" as foundation for overall formation and stress the existence of minor difference according to environment features. Therefore the dispersed community of "Xixi Institute" becomes an organic whole.

由明确几何体组合所构成的离散式聚合关系的"西溪学社"
"Xixi Institute", a dispersed community constituted by explicit geometry

"西溪学社"整体平面图
Masterplan of "Xixi Institute"

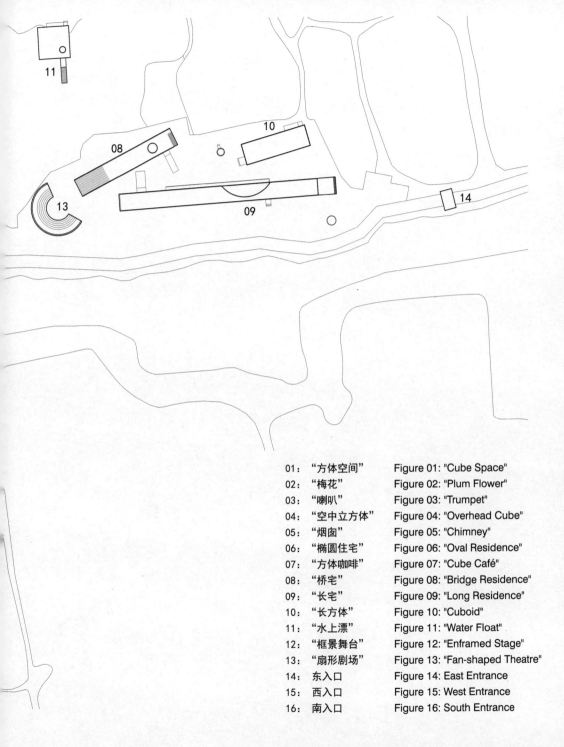

01：	"方体空间"	Figure 01: "Cube Space"
02：	"梅花"	Figure 02: "Plum Flower"
03：	"喇叭"	Figure 03: "Trumpet"
04：	"空中立方体"	Figure 04: "Overhead Cube"
05：	"烟囱"	Figure 05: "Chimney"
06：	"椭圆住宅"	Figure 06: "Oval Residence"
07：	"方体咖啡"	Figure 07: "Cube Café"
08：	"桥宅"	Figure 08: "Bridge Residence"
09：	"长宅"	Figure 09: "Long Residence"
10：	"长方体"	Figure 10: "Cuboid"
11：	"水上漂"	Figure 11: "Water Float"
12：	"框景舞台"	Figure 12: "Enframed Stage"
13：	"扇形剧场"	Figure 13: "Fan-shaped Theatre"
14：	东入口	Figure 14: East Entrance
15：	西入口	Figure 15: West Entrance
16：	南入口	Figure 16: South Entrance

构成"西溪学社"离散式会所的建筑
Architecture consist of "Xixi Institute"

1. "方体空间" "Cube Space"

建筑面积：416.22m²
total floor area：416.22m²

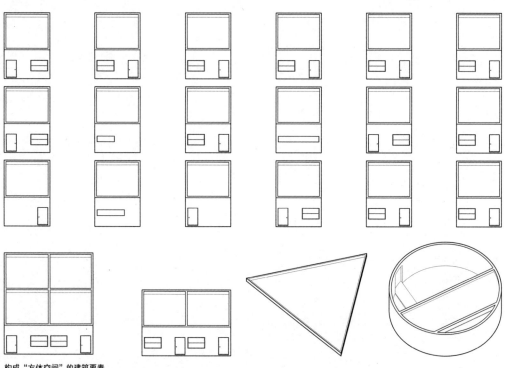

构成"方体空间"的建筑要素
Key architectural factors constituting "Cube Space"

"方体空间"这组建筑，在西溪学社中是一组小的旅馆建筑组合体。建筑总面积为416.22m²，由24个单元立方体所构成，每一个单元立方体房间都有单独的对外出入口，是一个独立的小住居的集合体。在这组建筑的东南侧有一个圆形的庭院，这是一个供管理者使用的公共区域。在设计这组建筑时，试图将传统聚落与现代建筑间进行直接转化，具体地讲，就是将作为设计师的"我"个人曾经调查过的中国传统聚落"日月山"的整体布局直接地转化和引用到该项目的设计之中。因为"日月山"的聚落本身是由一个个独立的住居所构成，而

The building of "Cube Space" is a small accommodation facility In "Xixi Institute". With building area of 416.22 square meters, it consists of 24 cube units, while each room, with its own door, is an independent residence. At the same time, on the southeast side of this building there is a round courtyard to serve public administrators. In designing this building, I attempted direct conversion. To be more specific, in designing this small inn, I tried to convert and quote the overall layout of "Riyueshan" Village, a traditional Chinese community I had personally surveyed. As that community itself consists of many independent residences, I tried to establish a similar independent feature when designing this small inn composed of cube space.The building itself employs cubes of 4m×4m×4m as the most basic spatial unit. Each

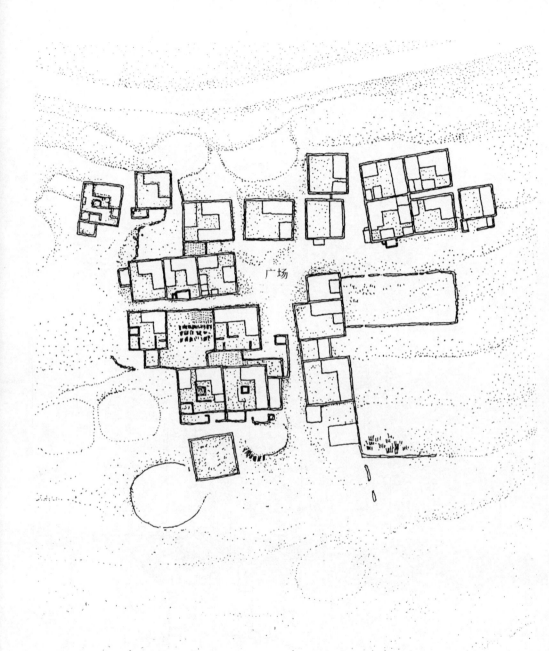

"日月山"聚落的测绘图
Survey map of "Riyueshan" Community

"方体空间"建筑轴测图
Axonometric drawing of "Cube Space"

对于新设计的由方体空间组合而成的小旅馆，也同样试图建立彼此之间的独立性，在这个层面上两者之间是契合的。这组建筑本身是以4m×4m×4m的立方体作为每一个最为基本的空间单元来使用，每一个单元本身是一个小客房，内部设置单元卫生间，从而使本来应该用室内走廊加以联结的旅馆性质的建筑成为一个可以直接对外的，拥有住居性质的聚落组合。而这种由立方体所形成的聚落本身，并没有直接引用原聚落的夯土材料，在采用一种新的语言进行构筑的同时，同样能唤起一种原初性感觉的回归。

unit is a small guest room with a washroom inside. A building that originally should have been connected by indoor corridor thus takes on community pattern that faces outside directly. By mixing the inherent concept of outdoor and indoor as well as residence and room that is defined by size, this practice helps future operation. This community formed by cubes, without directly using rammed earth as used in original community, adopts a new language in construction and arouses a sense of returned originality.

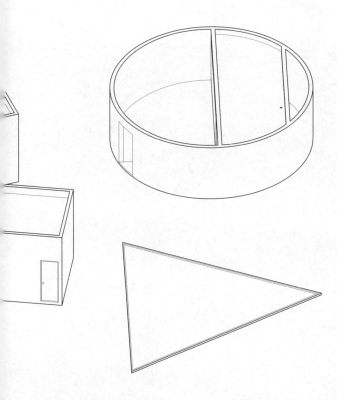

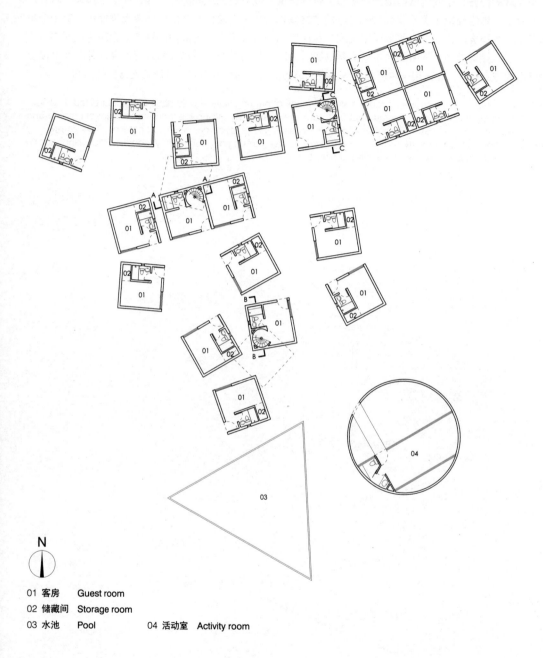

01 客房　　Guest room
02 储藏间　Storage room
03 水池　　Pool
04 活动室　Activity room

一层平面图　1st floor

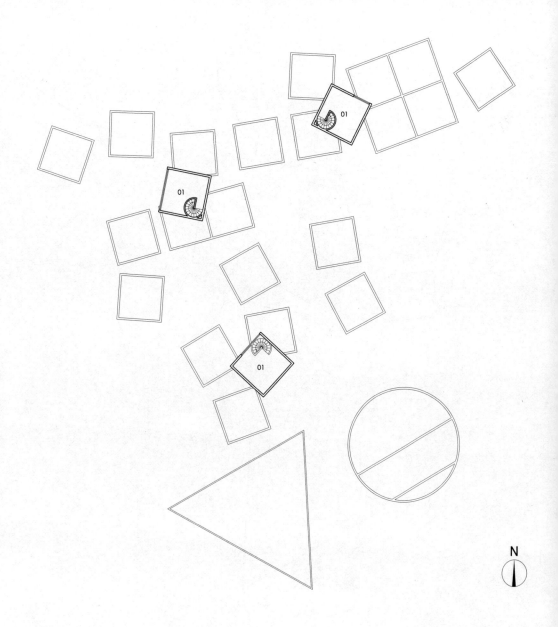

01 书房 Study room

二层平面图 2nd floor

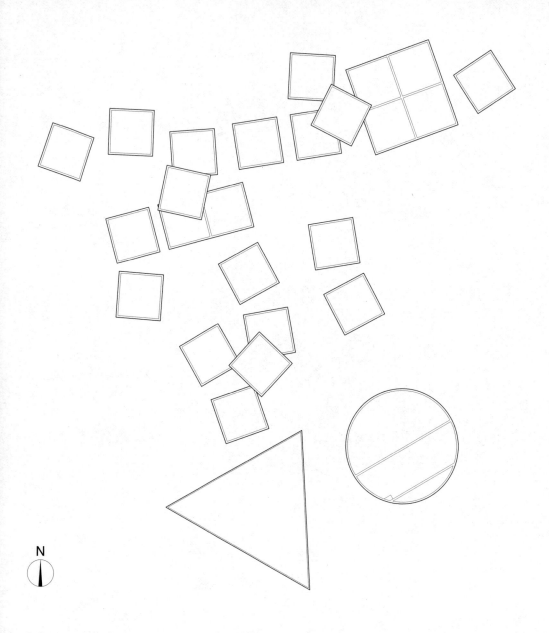

屋顶平面图 roof plan
0 1 5m

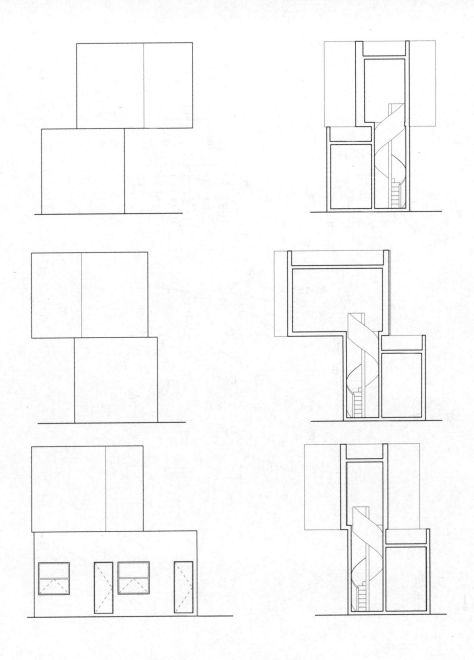

A1-5 南立面图	A1-5 south elevation	A-A 剖面图	A-A section
A1-19 南立面图	A1-19 south elevation	B-B 剖面图	B-B section
A1-13 南立面图	A1-13 south elevation	C-C 剖面图	C-C section

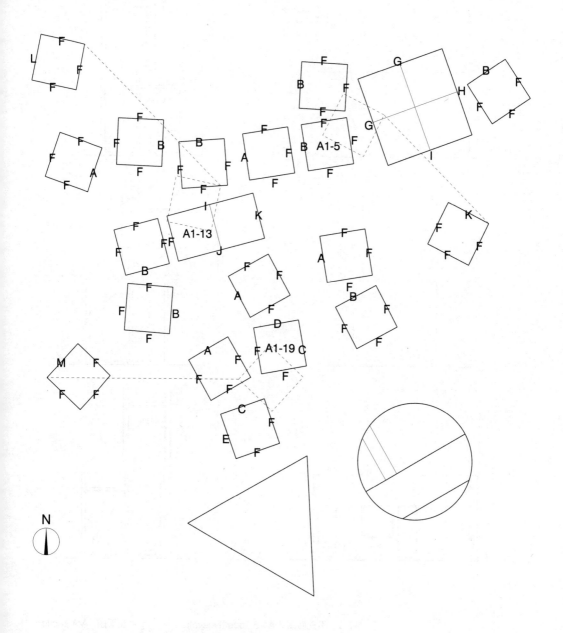

立面、剖面位置图 elevation&section map

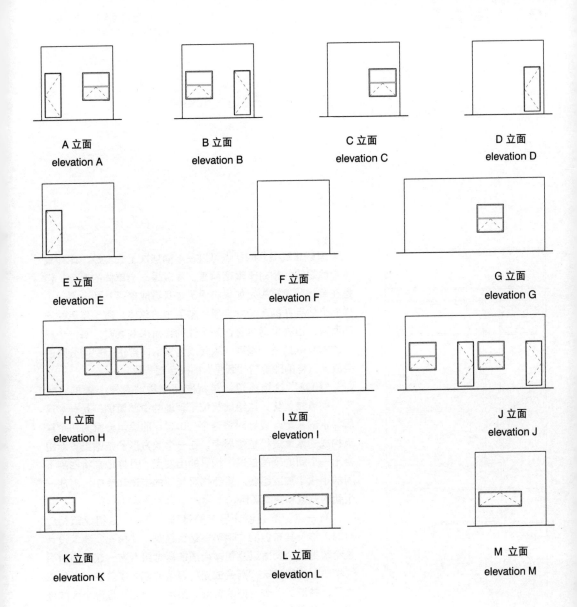

立面图 elevation

2. "梅花" "Plum Flower"

建筑面积：490.01m²
total floor area：490.01m²

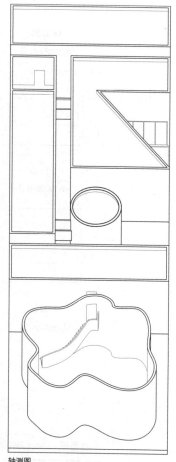

轴测图
Axonometric drawing of "Plum Flower"

  理解建筑概念的视点可以在不同层次上去穿越，比如聚落中的建筑，相对于聚落而言，建筑是一个单体的概念；但是在单体的建筑中，如果将建筑本身看成聚落，那么每一个作为个体存在的房子就会成为聚落中的单体。对于建筑的内部而言，当将房间本身作为个体的视角来观察时，每一个房子的本身就是一个聚落，家具成为聚落内部进行空间分隔的要素。这样的理解与"套匣"的概念相一致。

  "梅花"这组建筑，是试图将这种"套匣"般的"分节"视点物象化。这组建筑位于基地的中部西侧，是一个前后邻水的基地，设计时将这个490.01m²的建筑分解为四个建筑构成的聚落进行整体思考。在一个长方形的范围里重新组合出一个新的聚落是这个构思的出发点。具体地，将四种不同的小住宅加以组合，重新构筑起新的街道和巷院，形成一个新的领域和空间场所。

  有一点影响到设计思考的是与"梅花"这组建筑相应对的，位于其东侧的"喇叭"这组建筑。"梅花"建筑设计是希望把一个聚落以及整体街道的概念组合在一起。而对面"喇叭"这组建筑，则是试图将聚落的整体覆盖在一个屋檐之下，并形成一个内部的聚落。虽然"梅花"是一个大的建筑，但若仔细观察，里面还有四个小的建筑，同时在东侧还有一个梅花状的庭院。这些建筑构成了一个小的聚落，而这个聚落又构成了一个建筑。这样一个从不同层次上来进行思考和设计的视点可以让使用者体验到建筑中有街道，从街道还可以分别进入四个不同的小住宅里，从而形成一种"套匣"状的组合。

"梅花"
"Plum Flower"

"梅花"建筑中的梅花形态是一个徒手画的曲线,是试图对观念中的曲线和现实中的曲线之间的关系有所观察。这个梅花所构成的庭院实际上是一个封闭的小院子,它的产生是试图为未来的使用者提供一个思考的场所。使用者可以在这里喝茶、读书和发呆,给想象提供一个可以发散的空间场所。而这个封闭的空间并不是从室外可以随意出入的,它隶属于东侧的住居,同时也必须从住居卧室的二层才能通过楼梯来到这个庭院。

　　位于"梅花"建筑对面的"喇叭"建筑,从某种意义上与"梅花"建筑有着某种相同的思考出发点,即都是在探讨聚落空间的组合与再呈现问题,同时两个建筑都是试图在相同尺度的长方形中展开空间的序列和功能布局。所不同的是在相同的长方形里,"喇叭"是在室内进行了一个新的街道组织,并且这个街道是一个封闭的街道,确切地说是一个封闭的聚落。而"梅花"却是一个开敞的聚落,有室外街道的同时,还有室外的广场和庭园。

Architectural concept can be understood on different levels. For example, a building in a community is a single item relative to the community, but if the building itself is regarded as a community, then each house existing as an individual will become a single item in such a community. As for the inner of a building, when we observe a house itself as a single item, then each house itself is a community, with furniture becoming factors inside the community to divide and allocate space. Such understanding is consistent with the concept of "sheath".

The building of "Plum Flower" tries to materialize the perspective of "section" similar to "sheath". The building is located on the western part of the central base, bordering water area back and forth. This building of 490 square meters is considered during designing as a community consisting of four sections. The starting point of the design is to regroup a new community within the rectangular scope. To be specific, we combine four kinds of small residences to reconstruct new streets, alleys and courtyard, thus forming a new area and space.

One factor that affects designing consideration is the building named "Trumpet", which is a corresponding building on the east side of "Plum Flower". The architectural designing of "Plum Flower" aims to combine the concepts of a community and whole streets, while the opposite building named

"Trumpet" tries to cover the whole community under one roof and form an internal community. Although "Plum Flower" is a big building, actually with careful observation one may find that it contains four small buildings plus one courtyard of plum flower shape. All these buildings constitute a small community, while this community again forms a building. Such perspective based on different levels of consideration and design enables users to have the experience that a building contain streets within and streets lead to four different small residences, forming a combination pattern of "sheath".
The plum flower shape of this building is a curve by free hand drawing, an attempt to observe the relationship between curves in conception and reality. The courtyard with such a plum flower shape is actually an enclosed courtyard, trying to provide future users with a venue for contemplation. Here users can drink tea, read books and simply relax in trance, offering imagination a place to diverge and develop. This enclosed space is not randomly accessible from outside but subordinated to eastern residence, therefore one can reach this courtyard only by staircases from the bedroom on the second floor of the residence.
"Trumpet" building opposite to "Plum Flower" building has somewhat same design starting point in certain sense, which is to discuss the combination and re-display of community space. Meanwhile, two buildings both attempt to arrange spatial extension and functional layout within the rectangular scope of same size. The difference is that within the same rectangle, "Trumpet" arranges a new organization of street indoors and this street is an enclosed street or more accurately an enclosed community, while "Plum Flower" is an open community with outdoor streets as well as outdoor square and courtyard.

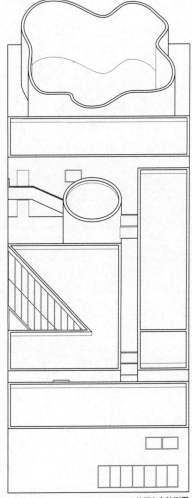

从西向东轴测图
Axonometric drawing from west to east

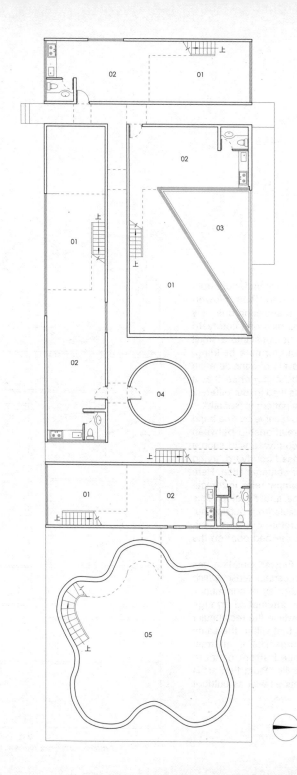

01 起居室	Sitting room
02 餐厅	Restaurant
03 水池	Pool
04 设备间	Equipment room
05 梅园	Plum garden

一层平面图 1st floor

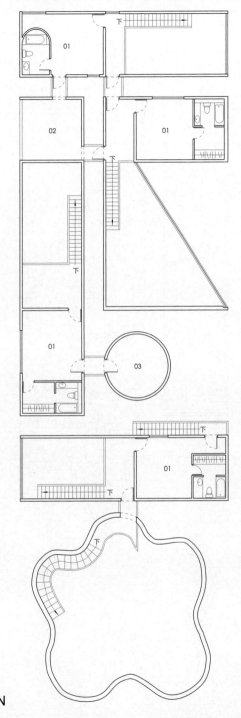

01 卧室　　Bedroom
02 内庭院　Inner courtyard
03 空调机房　Air-conditioning equipment room

二层平面图　2nd floor
0　1　　　5m

 N

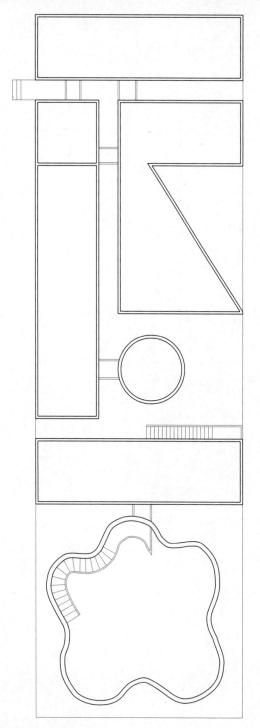

屋顶平面图 roof plan

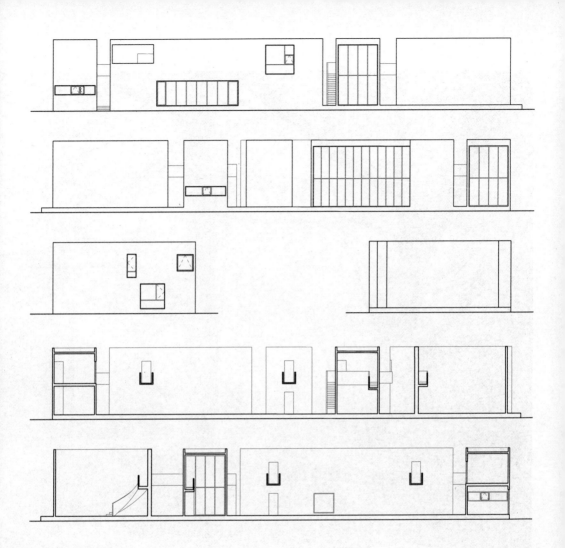
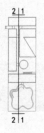

南立面图	south elevation
北立面图	north elevation
西、东立面图	west & east elevation
纵向剖面图1	longitudinal section 1
纵向剖面图2	longitudinal section 2

"梅花"建筑北立面
North elevation of "Plum Flower" building

"梅花"建筑与环境
"Plum Flower" building and surroudings

"梅花"庭院
"Plum Flower" Courtyard

"梅花"建筑北侧风景
North Scenes of "Plum Flower"

"梅花"东北角
North-east corner of "Plum Flower"

"梅花"庭院内楼梯
Staircase of "Plum Flower" Courtyard

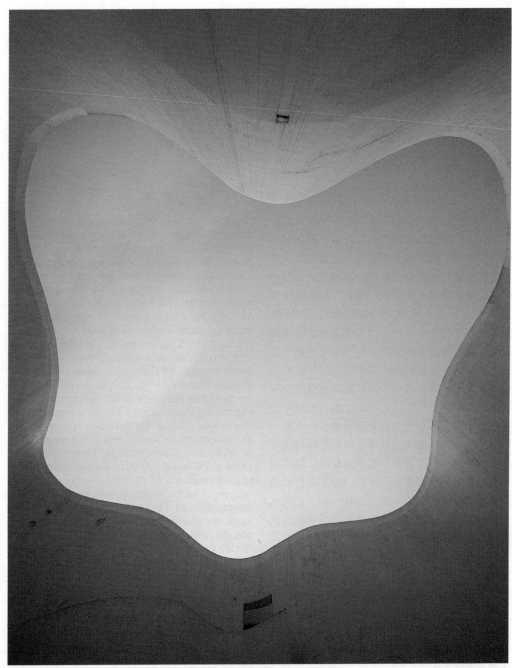

"梅花"庭院上空
Above "Plum Flower" Courtyard

3. "喇叭" "Trumpet"

建筑面积：661.85m²
total floor area: 661.85m²

"喇叭"在位置上是与"梅花"相互对应的建筑。"喇叭"建筑的主入口放在基地的北侧。"喇叭"的体量内部是一个公共空间，可作为居住者的交流场所使用。从喇叭的开口方向可以欣赏到对面摆放的"梅花"。这组建筑面积为661.85m²，设有一个中心庭院，围合庭院的建筑内部是一个封闭的街道系统。将聚落的街道空间进行室内转化，是希望与西侧"梅花"建筑的室外化街道布局产生对照和比较。进入建筑的一层，是一个环状的内部空间，环状街道上空布置有飘浮状态的住居。一层环状街道是所有二层居民的公共活动场所，也是一个彼此之间进行交流的空间。一层环状街道中有分别能够上到二层各个客房的楼梯。飘浮在二层的这些客房朝向庭院一侧，分别有可以通到中央庭院里的梯段。通过这样一种对空间出入口的控制，使整体的建筑路径形成一种迷宫式的空间布局。此外在"喇叭"这栋建筑中，还设计有两个不同的路径系统，除上述在一层室内设置的内部街道外，从室外通过西南侧的旋转楼梯进入中心庭院的路径是使中心庭院具有下沉式"窑洞"特征的措施。将相同标高的空间，通过路径的设置而改变其空间的高程感受，是一种魔术式穿越路径的设置，更强调了人的感受性主体的存在。

"Trumpet", whose main entrance is on the north side of the base, is a building corresponding to "Plum Flower" in terms of location. Inside the building there is a public space for exchange and communication among residents. From the opening of the trumpet one can enjoy the view of opposite "Plum Flower". This building, with built-up area of 661.85 square meters, has one central courtyard. Inside the building of enclosed courtyard there is an enclosed street system. By transforming the space of community streets into indoor space, we hope to create contrast and comparison against the outdoor street of "Plum Flower" building on the west. The first storey of this building is a circular internal space, with floating style residence above the circular street. The circular street on the first storey is the venue for all residents on the second storey to conduct public activities and communication. From the circular street on the first storey there are staircases leading to each residence on the second storey. As to the residences on the floating second storey, they have staircases leading to central courtyard on the side facing the courtyard. By controlling space entrance and exit, the whole building path forms a labyrinth-like layout pattern. Moreover, there are two different path systems in "Trumpet". In addition to the internal street indoors on the first storey, the other path, leading to central courtyard from outdoors through the spiral staircase on southwest side, attaches the feature of sunken "cave dwelling" to the central courtyard. For space of same elevation, the designer manages to change the feeling of altitude by magical path setup, which emphasizes the existence of people as the subject in perception.

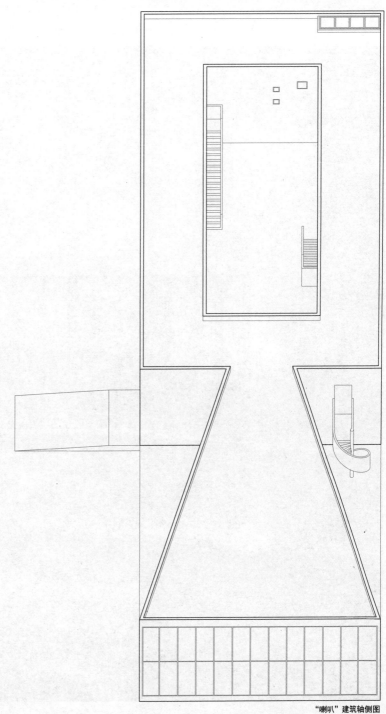

"喇叭"建筑轴侧图
Axonometric drawing of "Trumpet" building

"喇叭"西侧
West side of "Trumpet"

"喇叭"南侧
South side of "Trumpet"

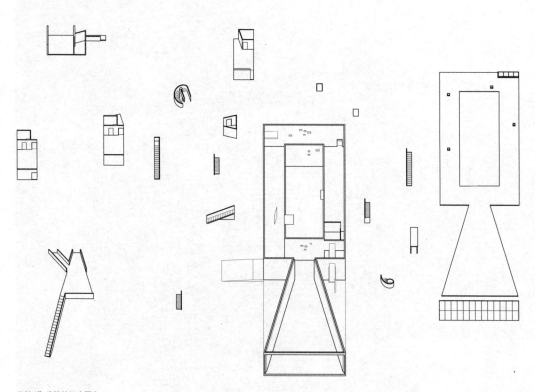

"喇叭"建筑的组合要素
Combination factors of "Trumpet" building

被转换了高程之后的中心庭院
Central courtyard after elevation transformation

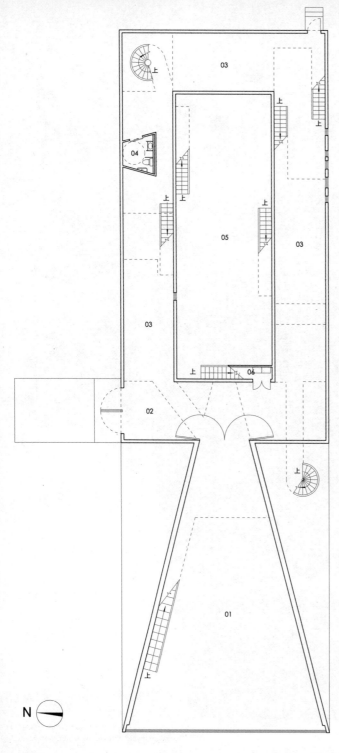

01	活动大厅	Activity hall
02	门厅	Foyer
03	起居室	Sitting room
04	残疾人卫生间	Washroom for the disabled
05	内庭院	Inner courtyard
06	设备间	Equipment room

一层平面图 1st floor

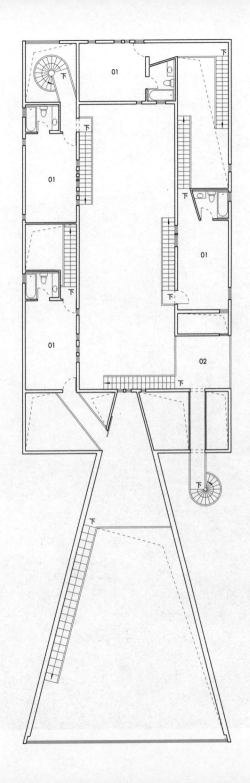

01 卧室　Bedroom
02 起居室　Sitting room

二层平面图　2nd floor

"喇叭"南立面
South elevation of "Trumpet"

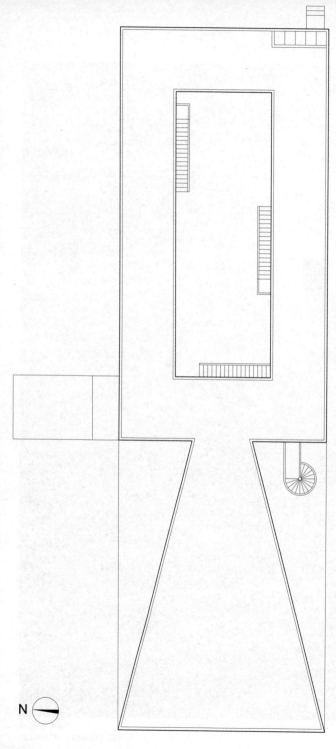

屋顶平面图 roof plan

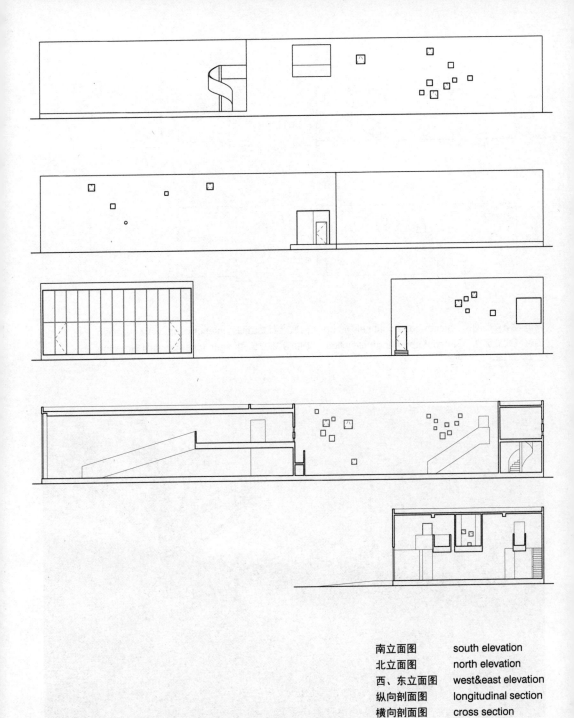

南立面图	south elevation
北立面图	north elevation
西、东立面图	west & east elevation
纵向剖面图	longitudinal section
横向剖面图	cross section

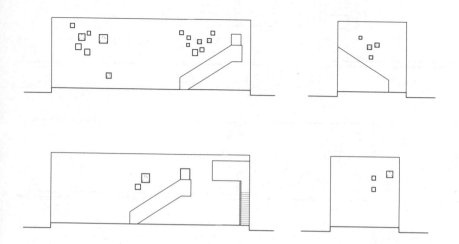

内庭院南立面图　interior court south elevation　　内庭院西立面图　interior court west elevation
内庭院北立面图　interior court north elevation　　内庭院东立面图　interior court east elevation

0　1　　5m

"喇叭"与环境
"Trumpet" and surroudings

"喇叭"东侧
East side of "Trumpet"

4. "空中立方体" "Overhead Cube"

建筑面积：109.00m²
total floor area：109.00m²

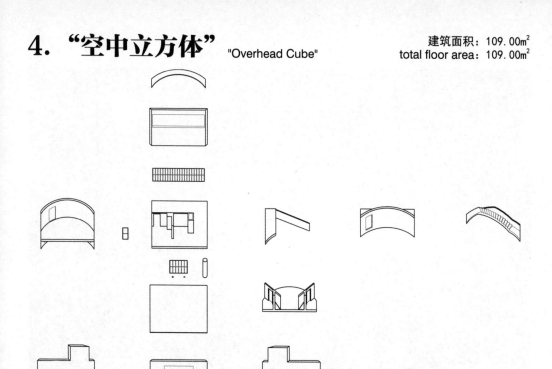

构成空中立方体建筑的要素
Architectural factors constituting "Overhead Cube"

在"西溪学社"的建筑群中，"空中立方体"是园区内最小的一栋建筑，该建筑位于整个基地的西侧入口，原本是为管理者设置的办公用房，所以设计时尽可能地缩小用地面积，也正是因为如此，房子的面积也设计得非常小，总体建筑面积只有109m²。

这个建筑尽管只有109m²，但实际上有四层楼高，目的是能使该建筑成为园区内的制高点。该建筑的一层是一个开敞空间，平时管理者可在一层开敞的亭台空间中值班。二层是一个小的起居室，并有厨房等可以满足生活最基本的功能性需求的配置。从二层起居室有一个

Amid the architectural group of "Xixi Institute", "Overhead Cube" is the smallest. Located at the west entrance to the base, it used to be office building for administration, resulting in minimum land area and room space in design. Its total built-up area is only 109 square meters.

With only 109 square meters, the building actually has four storeys, becoming the commanding high point in the area. The first storey is wide open space for administrators to work on duty. The second storey is a small living room equipped with kitchen to meet most basic life needs. There is a small staircase from the living room on the second storey to the third storey, where a small bedroom is ready for administrators to rest. On one side of the bedroom there is a staircase to roof, which is

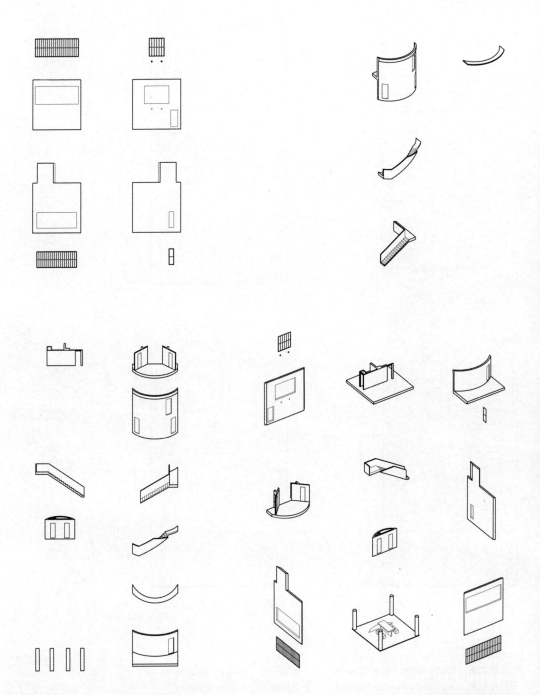

构成空中立方体建筑的要素
Architectural factors constituting "Overhead Cube"

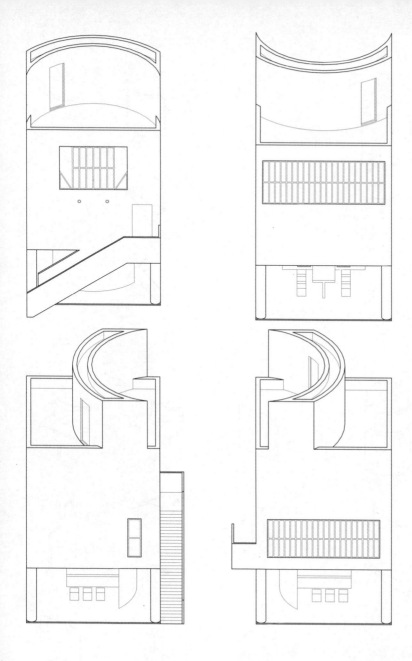

| 正北轴测图 | north axonometric drawing | 正南轴测图 | south elevation |
| 正东轴测图 | east axonometric drawing | 正西轴测图 | north elevation |

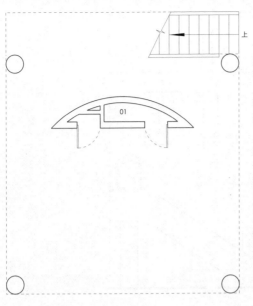
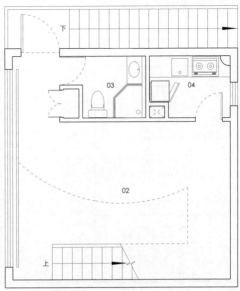
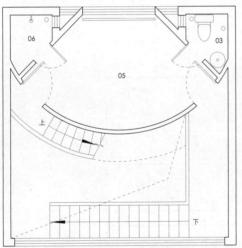
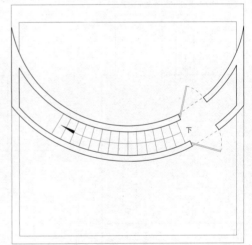

01	设备间	Equipment room	04 厨房	Kitchen
02	起居室	Sitting room	05 卧室	Bedroom
03	卫生间	Washroom	06 浴室	Bathroom

一层平面图 1st floor　　二层平面图 2nd floor
三层平面图 3rd floor　　屋顶平面图 roof plan

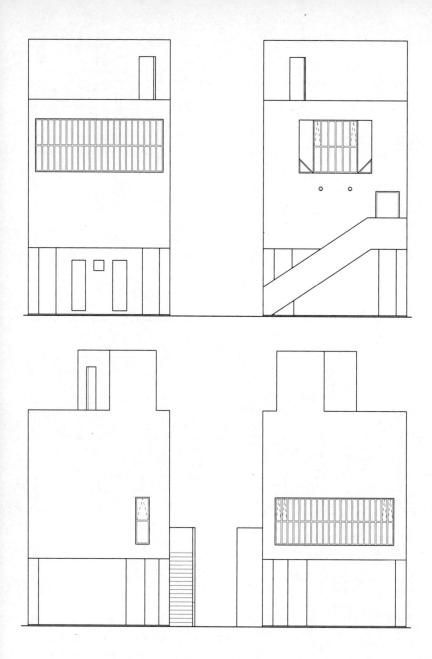

南立面图　south elevation　　北立面图　north elevation
东立面图　east elevation　　　西立面图　west elevation

小楼梯可以进入三层，那里有一个小的卧室，供管理者休息使用。在这个卧室的一侧还有楼梯可以通往屋顶，这个屋顶的高度在整个园区里面是一个制高点，管理者可以站在这个平台上对整个园区进行管理和监视。同时由于这个建筑的体量只有6m×6m，从而获得整个园区中的这样一种小而高的体量状态。

a commanding height where administrators can manage and monitor the whole area. As this building is only 6m x 6m in mass, it acquires a small but high status in the whole area.

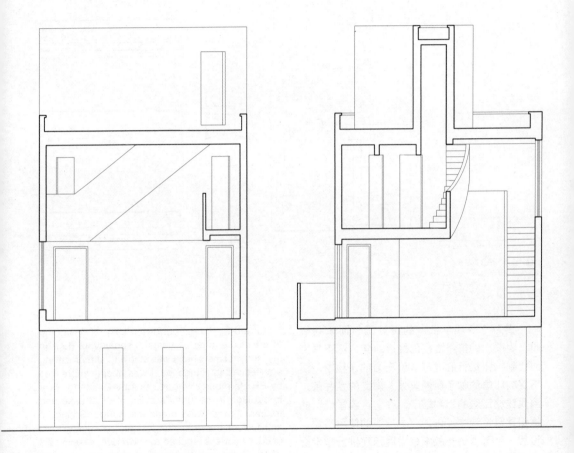

横向剖面图 cross section　纵向剖面图 longitudinal section

5. "烟囱" "Chimney"

建筑面积：772.50m²
total floor area：772.50m²

构成"烟囱"的建筑要素
Architectural factors constituting "Chimney"

我为"西溪学社"的设计工作而重到杭州，距前次的杭州之旅已经近21年。记得学生时代到杭州进行的建筑认识实习的课程中，留下深刻印象的除了西湖美景、虎跑的龙井茶，当属钱塘江边的六和塔了。在"西溪学社"的这组建筑设计中，依据"学社"的使用需要，设置一个可以进行教学成果展示和进行学术交流的场所。于是"烟囱"便应运而生。这个设想开始是试图与现有场地中现存的几个砖砌的烟囱遗构之间形成关照。然而实际上，当建筑完成之后，却勾起我对于六和塔的联想，这个原本只是作为烟囱而思考的建筑本身无意间却

I returned to Hangzhou for the sake of designing "Xixi Institute" after 21 years of separation. Back in my student time when I was studying and practicing architecture in Hangzhou, I was deeply impressed by Six Harmonies Pagoda nearby Qiantang River in addition to beautiful scenes of West Lake and fragrant Dragon Well green tea in Tiger Spring. In architectural design of "Xixi Institute", according to its need, a venue is required to showcase achievement of teaching and learning as well as carry out academic exchange."Chimney" thus takes shape. This conception at the beginning attempts to form corresponding relationship with several remnant brick chimneys on the site, but actually after its completion, it arouses my associated memory of Six Harmonies Pagoda. This building, merely regarded

从西南侧看"烟囱"的整体
Whole view of "Chimney" from south-west side

从北侧看"烟囱"的整体
Whole view of "Chimney" from north side

具有了文化的地域性指向。

　　这个充满不经意而与六和塔"巧合"的塔身，原本是想设置一个充满光的"烟囱"，将黑暗从室内"拔出"。该建筑的内部由两层构成，一层向地下沉降了1.5m，当人们从室外标高进入室内后，对整个大厅视线本身并不是一个仰视的状态，而是产生一个可以上下进行选择的视线。人的视点处于一个空间的中间位置而使空间显得平和。这样做的目的一方面是为了降低整体建筑的体量，另一方面是可使位于室内一层的人能从非日常性的视点透过通宽的横向玻璃，重新观察和审视湿地的日常性风景并获得非日常性的感受。

　　从建筑的南侧进入，有一个向上和向下的坡道，建筑的内部通过直跑楼梯和螺旋楼梯构成一个完整的回路。高塔将光送到大厅中。整个空间内部的柱子构成的柱阵形成与湿地场景相呼应的树林感觉。一层与二层的窗户采用连续的条窗，希望造成类似中国画卷轴山水长卷般的连续性展示。此外，考虑到未来的经营问题，该项目在设计时还力图建成满足未来多种可能性的使用场景，建筑本身不仅可以作为一个小的展示空间，同时也不排除将其改造为会所来使用。譬如，在二层预留了设备管道的位置，未来可以在此改造，做成一间间客房；还有一种就是把该建筑做成小的餐饮场所。这几种使用的可能性都在设计时作了预设处理。

as a chimney at first, unintentionally possesses cultural and regional orientation.
This building that "accidentally coincides" with Six Harmonies Pagoda originally intended to become a "chimney" full of light that "extracts" darkness from indoors. This building is composed of two storeys indoors. The first storey sinks 1.5 meters below ground. When people enters indoors from outside, they can choose visual line up and down, as the visual line inside the hall is not in a look-up status. The point of sight is in the central part of space, making the space appear mild. The purpose of doing so is not only to reduce the total architectural mass but also enable people on the first storey indoors to observe and examine the daily wetland scenes again, from unconventional viewpoint and through horizontal glass, to obtain unconventional experience.

When one enters the building from the south side, there is one slope upward and downward. A complete circuit is formed inside the building by straight staircase and spiral staircase. A high tower brings light into the hall. The column pattern composed by pillars inside the space produces a feeling of woods corresponding to wetland scenes. Windows of the first and second storey are continuous slots, trying to achieve the continuous display effect of Chinese landscape scrolls. Besides, considering future operation, the project also takes possible future scenarios into consideration during designing. The building itself can be used not only as a small showroom but also as a club after renovation. For example, on the second storey places for equipment pipes are reserved for future renovation into guestrooms. Another way is to turn this building into small food and beverage facility. All these possible usages are prepared during designing.

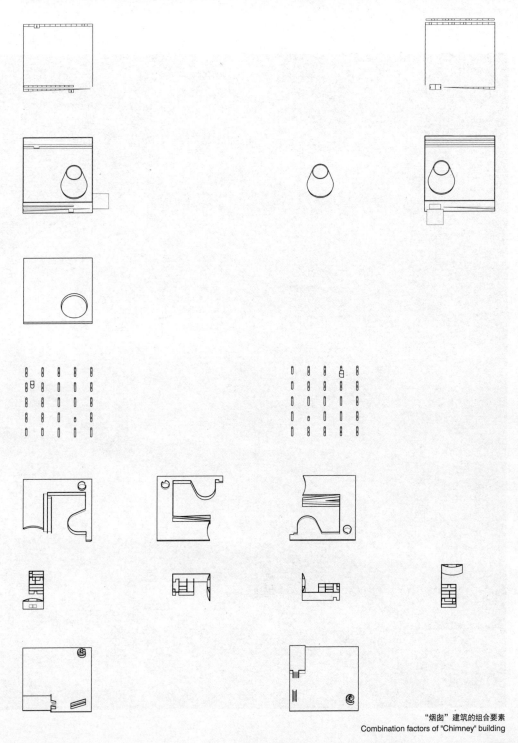

"烟囱"建筑的组合要素
Combination factors of "Chimney" building

"烟囱"室内
Indoors of "Chimney"

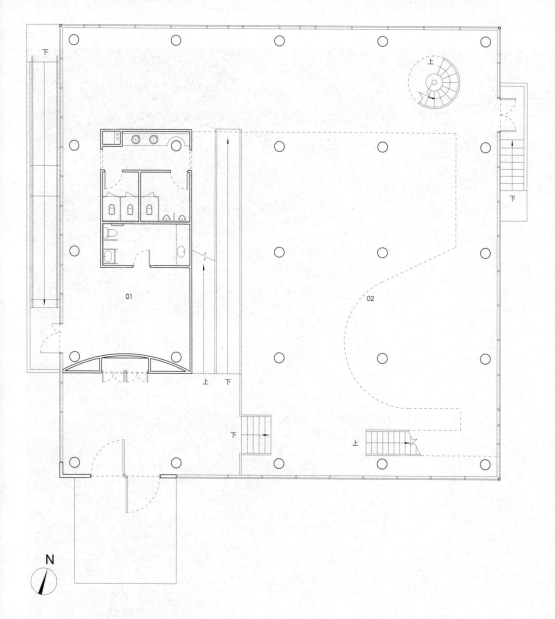

01 电加热厨房　Electrical heating kitchen
02 餐饮区　　　Catering area

一层平面图　1st floor

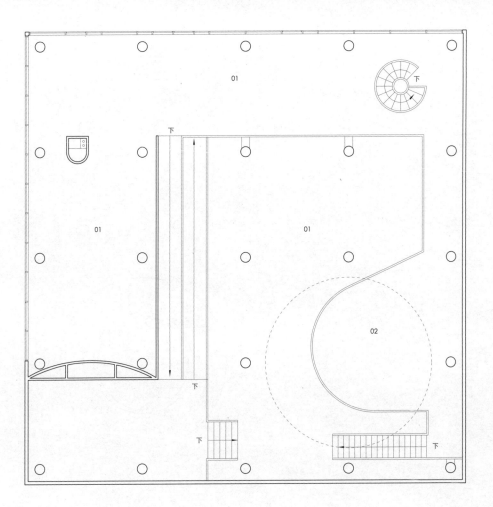

01 餐饮区　Catering area
02 休息区　Rest area

二层平面图　2nd floor

屋顶平面图 roof plan

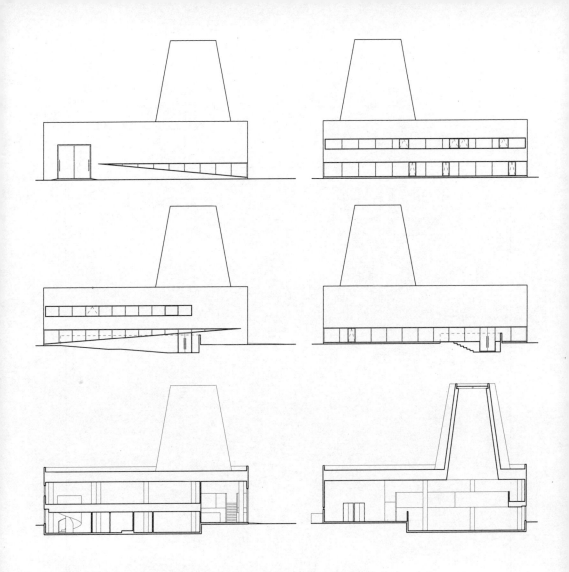

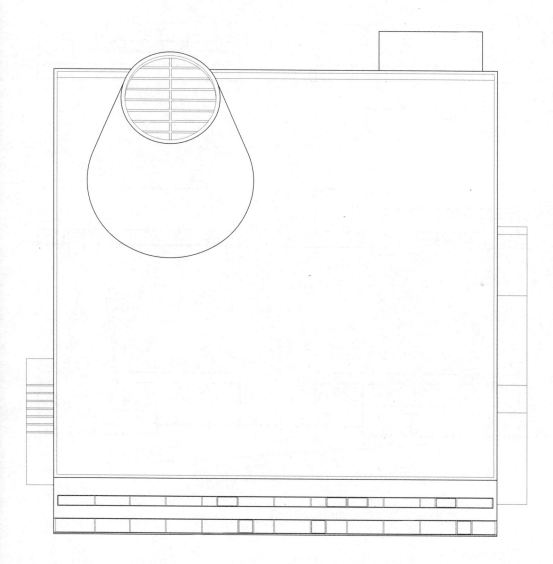

正北轴测图
north axonometric drawing

"烟囱"东立面
East elevation of "Chimney"

"烟囱"建筑的东南角风景
South-east corner of "Chimney" building

6. "椭圆住宅" "Oval Residence"

建筑面积：251.60m²
total floor area：251.60m²

椭圆并不是一个欧几里得几何学意义的存在，从希腊的欧几里得几何学到罗马时期的椭圆的产生，几何学的变化是否一定带来建筑的进步问题实际上是我一直思考、关心、无法理清并充满疑问的问题。椭圆的性格附和了从"进化论"的视角来理解几何学的要求。这个椭圆住宅，正是试图让自己"进步"而采取的措施。本来这所房子，开始只是希望这个椭圆形的弧墙仅仅作为围墙而存在，房子存在于围墙之中，并且房子与围墙之间是彼此完全脱离的状态。曾将其命名为"圆中园"，取在一个椭圆的围墙里面包含一个小的园子的意向，并希望在椭圆里面围合出几个不同形状的庭院。

目前这个建筑为251.6m²，整个建筑为一层，南侧紧邻基地中的东西主干道。建筑从入口开始设置了一个街巷，形成一个过渡空间，入口的另一侧一个弧形的街巷过渡到北侧的庭院之中。在北侧庭院的一侧有一个小坡道上升到通往二层并悬挑于北侧的空中小舞台上，在这个舞台上，可以观赏到湿地北侧的风景。建筑中的起居室和餐厅是一个模糊界面的存在，这两部分弧形墙面的微妙变化，让空间的界面模糊甚至消失，使观者无法察觉空间的真实尺度，从而使房子获得空间层面的意义表达。

Oval is not an existence in Euclid geometry. I am always not only involved and interested but also confused and puzzle by whether or not the change in geometry definitely brings architectural progress, from Euclid geometry in ancient Greece to the birth of oval in Roman Empire. The feature of oval complies with the need to understand geometry from the perspective of "evolution". This Oval Residence is exactly an action attempting to make myself "progress". At first I expected that the oval wall existed only as an enclosing wall, with the house inside the wall but totally separated from the wall. It was once named as "Garden within Oval", meaning a small garden, hopefully with several courtyards of different shapes, to be contained inside an oval enclosing wall.

Currently the built-up area of this one-storey building is 251.6 square meters, bordering the east-west main road in the base on the south side. From the entrance of this building, a street is arranged to form transition space. Meanwhile, on the other side of the entrance there is an arc-shaped street transiting to the courtyard on the north. From the side of north courtyard there is a small slope leading to the second storey and further to the overhanging stage on the north side. On this stage one can enjoy scenes on the north side of the wetland. Living room and dinning room in this building exist on a vague interface, as the subtle change of arc-shaped walls in these two parts make spatial interface ambiguous and even disappearing so that observers cannot perceive the real spatial size, enabling the house to express meaning on spatial level.

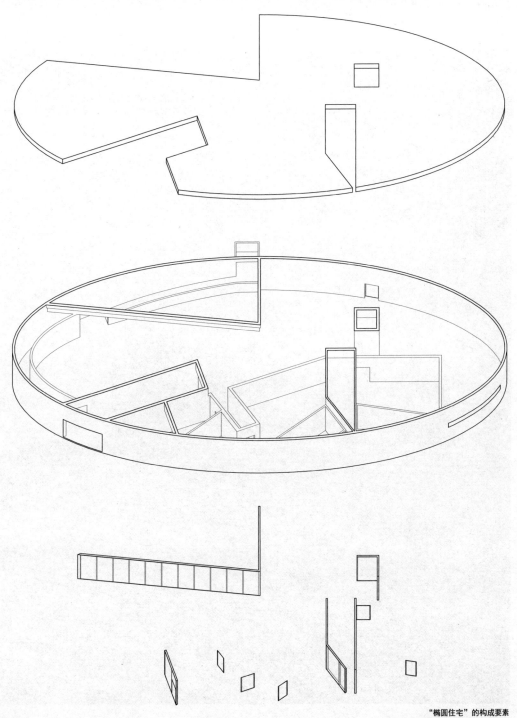

"椭圆住宅"的构成要素
Factors constituting "Oval Residence"

"椭圆住宅"东南侧风景
South-east side of "Oval Residence"

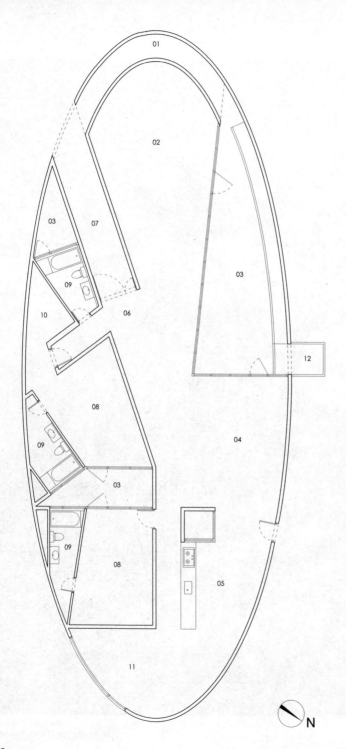

01	走廊	Passageway
02	多功能厅	Multifunctional hall
03	内庭院	Inner courtyard
04	起居室	Sitting room
05	厨房	Kitchen
06	门厅	Foyer
07	门廊	Porch
08	卧室	Bedroom
09	卫生间	Washroom
10	衣帽间	Cloakroom
11	餐厅	Diningroom
12	阳台	Balcony

一层平面图 1st floor

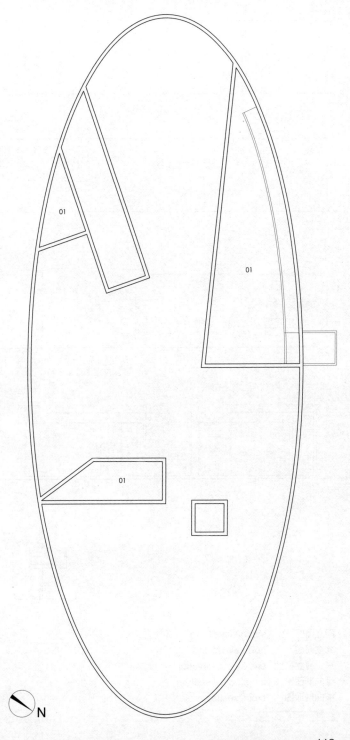

01 内庭院 Inner courtyard

屋顶平面图 roof plan
0 1   5m

N

113

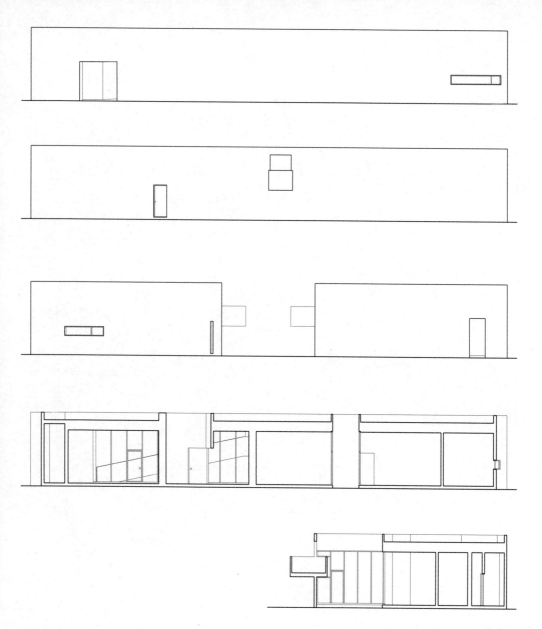

南立面图	south elevation
北立面图	north elevation
东、西立面图	east&west elevation
纵向剖面图	longitudinal section
横向剖面图	cross section

"椭圆住宅"西北侧
North-west side of "Oval Residence"

"橢圓住宅"
"Oval Residence"

"椭圆住宅"与周围建筑的体块组合关系
Mass combination relationship between "Oval Residence" and nearby buildings

7. "方体咖啡" "Cube Café"

建筑面积：9.00m²
total floor area：9.00m²

在北侧紧邻"椭圆住宅"，南侧靠近园区东西向的主干道路旁，有一个3m见方的立方体，这个只有9m²的立方体是整个园区中最小的建筑，同时也是最小的咖啡厅。

这个咖啡厅并不是一个能够供使用者进入的咖啡厅，而仅仅是为咖啡厅提供了一个可以调制咖啡和准备食品的空间场所。由于这个建筑的尺度很小，仅能作为一个咖啡厅的厨房使用，所以在这个厨房中仅安放有必要的上下水，相关电器和厨具设备，在此可以完成制作茶点和咖啡的工作。咖啡厅的客人区域设置在椭圆建筑南侧与道路北侧之间的空地上，是一个露天的咖啡厅。露天咖啡厅的地面上铺有细碎的石子。在石子地面上设置有桌椅，形成了"西溪学社"的街边咖啡角。

这个9m²的建筑西侧，安放一个巨大的门扇，这个门扇本身可以作为一个巨大的广告牌，同时还是一个限定空间的墙壁。白天经营的时候将门扇打开，整个建筑的方向朝向露天的咖啡角，从而使这个露天的外部空间拥有一个视觉中心的意义。当晚上关闭时，建筑本身成为路边的方体，仅占据极少的用地资源。这个咖啡厅的设计是对如何利用极小资源来获得最大空间支配感的尝试。

There is a 3m×3m cube whose north side is close to "Oval Residence" and south side is close to east-west main road of Xixi Wetland. With only 9 square meters, it is not only the smallest building in Xixi Wetland but also the smallest café.
Instead of a place for customers to enter, this café is only for preparing coffee and food. Due to its small size, it serves only as a kitchen of a café. It is equipped with necessary plumbing, electric appliances and kitchen equipment to complete preparation of snacks, tea and coffee. Customers can stay outdoors between the south side of Oval Residence and north side of the street. With tiny pebbles on the ground of the outdoor café as well as chairs and tables, a roadside coffee corner takes shape in "Xixi Institute".
On the west side of this building of 9 square meters, there is a huge door. This door can also serve as a huge advertising board as well as a wall to define space. In the daytime the door is open and the whole building faces the outdoor coffee corner, therefore the outdoor space has a visual center. At night the door is closed and the building becomes a roadside cube, occupying very little land resources. The design of this café is an attempt to use minimum resources for maximum space governance.

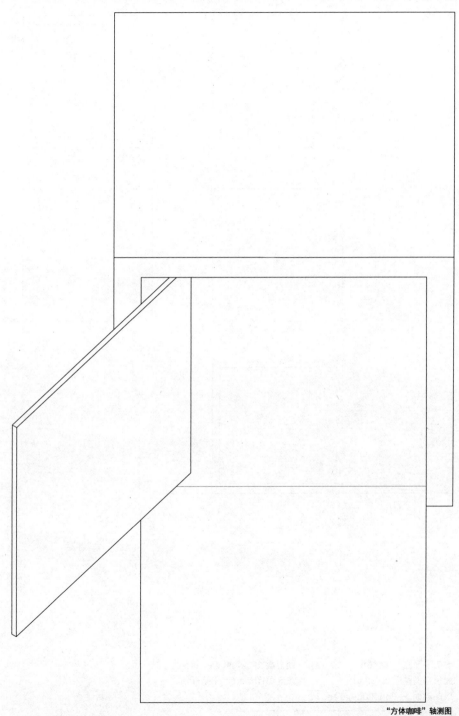

"方体咖啡"轴测图
Axonometric drawing of "Cube Café"

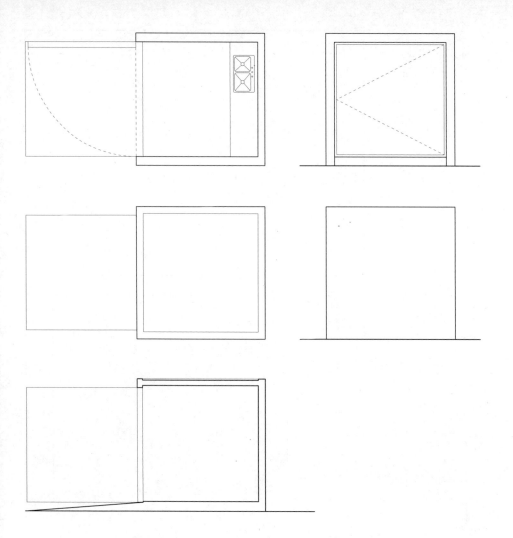

一层平面图	1st floor
屋顶平面图	roof plan
纵向剖面图	longitudinal section
西立面图	west elevation
东立面图	east elevation

街角"方体咖啡"
"Cube Café" at street corner

8. "桥宅" "Bridge Residence"

建筑面积：172.53m²
total floor area：172.53m²

江南一带的水乡风景中，桥是一个重要的要素，当人们在水乡中行走，桥的存在让原本普通的平地变得富有起伏和变化，尽管这种桥的存在是功能性的，但是这种功能性的本身却同时带来了特有的空间体验及丰富的记忆感受。"桥宅"位于"西溪学社"的东区，建筑面积为172.53m²。建筑西侧是一个直通屋顶的大台阶，与位于其西侧的圆形小剧场形成呼应。这个大台阶所构成的形态与江南一带聚落中桥的形态相应和并构成相互之间在空间层面上的链接。该建筑的东侧有一个小楼梯，与西侧的大台阶共同构成连接屋顶平台的回路关系。人们在这里上下来往的行为本身，恰恰是一个非常重要的地域性的体验方式。实际上地域性的再提示本身并不一定仅仅停留在运用当地材料和形式这样简单的表象层次，因为当今建筑材料的普遍性早已均质。而建立在生活上的行为与空间感受的呼应，或许会令使用者产生一种富有地域性的空间场景和意识的联想。这个建筑的屋顶下方是一个大空间工作室，大空间的一侧有生活必须的卧室和厨房。建筑用的空调室外机等机电设备放在屋顶上，同时作为围合这些设备的墙体本身成为这个建筑屋顶的一个构成要素。这个屋顶平台也成为一个可举行交流活动的公共场所。可以说，"桥宅"投射着设计者脑中江南水乡的观念性风景。

Bridge is an important factor in Jiangnan region (roughly the region in southeast China around Yangtze River) of rivers and lakes. When people are walking in such areas, the existence of bridge makes ordinary flat ground full of fluctuation and change. Although the existence of this kind of bridges is functional, this functionality itself also brings unique spatial experience and rich memory. "Bridge Residence" is located on the east zone of "Xixi Institute" with built-up area of 172.53 square meters. On the west side of the building leading directly to its roof there is a big staircase, which corresponds to a small amphitheatre on the west. The pattern formed by this big staircase corresponds to the pattern of bridge in Jiangnan communities, constituting linkage on spatial level between each other. On the east side of the building there is a small staircase, forming a loop to the roof together with the big staircase on the west side. The behavior of climbing up and down is exactly a very important way to experience regional features. Actually the presentation of regional feature again does not solely rest on simple and superficial level of using and adopting local materials and forms, as currently the universality of building materials are homogenized for a long time. However, the correlation based on living behavior and spatial experience may arouse associated thoughts full of regional spatial scenes and consciousness from users. Under the roof there is a spacious studio, equipped with life necessities such as bedroom and kitchen on one side. Mechanic and electric appliances such as air conditioning outdoor units are placed on the roof, enclosed by walls that form one factor of the building roof. This rooftop terrace also becomes a public venue to hold exchange activities. It is said that "Bridge Residence" reflects conceptual scenes of Jiangnan region of rivers and lakes in the mind of the designer.

"桥宅"
"Bridge Residence"

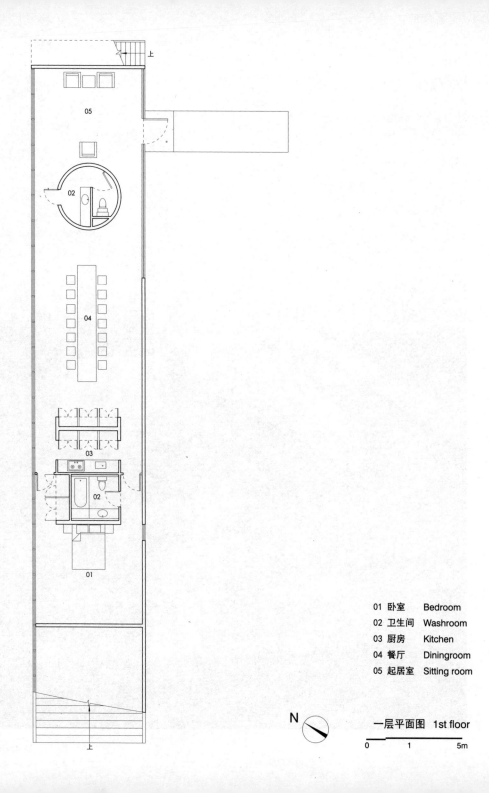

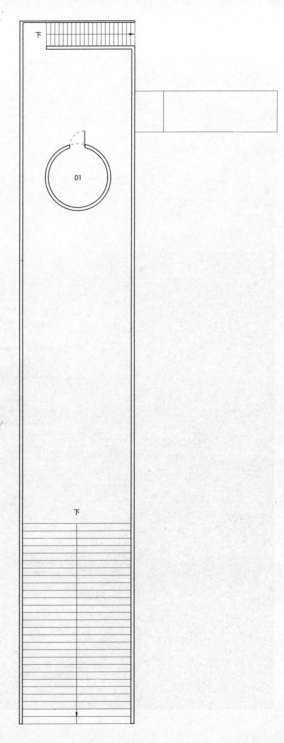

01 设备间 Equipment room

屋顶平面图 roof plan

"桥宅"风景
Scenes of "Bridge Residence"

"桥宅"屋顶平台
Roof terrace of "Bridge Residence"

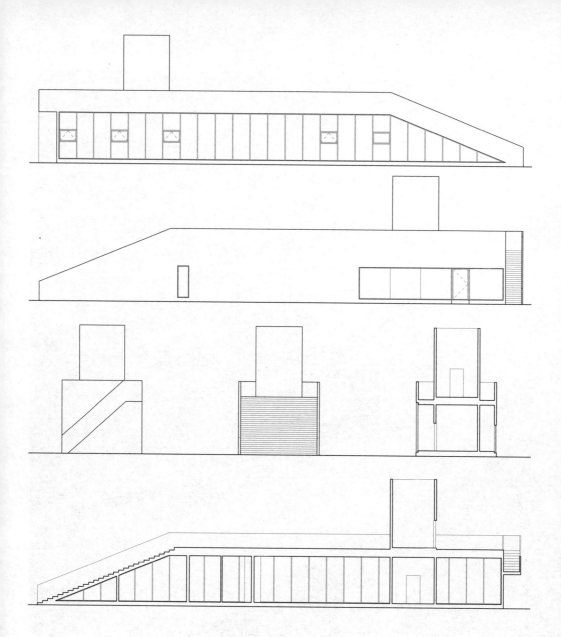

北立面图　　　　　　　　　north elevation
南立面图　　　　　　　　　south elevation
东、西立面图、横向剖面图　east&west elevation&cross section
纵向剖面图　　　　　　　　longitudinal section

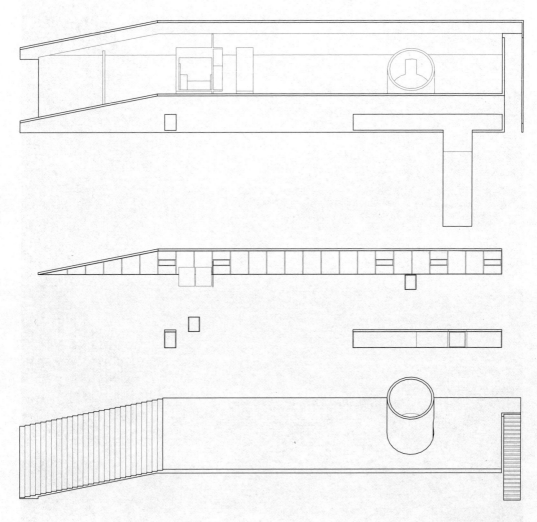

"桥宅"建筑的构成要素
Factors constituting of "Bridge Residence"

"桥宅"与周围建筑的体块关系
Mass relationship between "Bridge Residence" and nearby buildings

"桥宅"东侧
East side of "Bridge Residence"

9. "长宅" "Long Residence"

建筑面积：579.50m²
total floor area：579.50m²

"长宅"位于"西溪学社"园区的东侧，"桥宅"建筑的南侧，与"桥宅"和"长方体"两个建筑共同围合成一个梯形的室外广场。"长宅"建筑为两层，建筑面积为579.5m²。由于该建筑是园区中最长的建筑，总长度为66m，因而被命名为"长宅"。该建筑在设计之初，南侧的玻璃幕墙是不存在的。原本是试图在一个狭长的体块中重新塑造聚落中广场和街道的关系。并由于"西溪学社"园区中的"梅花"建筑是一个传统意义上的街巷式聚落布局，"喇叭"是把聚落的街巷关系覆盖到室内，"长宅"这个建筑则是试图将一个聚落的要素横向展开并框在一个框子之中，构成如同"街村"型的聚落布局。"长宅"建筑的一层对外展开，有不同的楼梯通向安排在二层的各自空间不同的小住宅。这些变化着并漂浮在街巷上空作为小住宅而存在的盒子，构成了单纯体块中的丰富性。当使用者在上下楼梯之时，他们可以并能够感受到面向南部的湿地景观。不过这样的设想事实上有些超出日常标准。为了让这个设计具象和符合日常生活，最终决定在这个半开敞的"框景城市"外部加上一层玻璃幕墙。"长宅"建筑的北侧同样是一个景色优美的湖区，为此在建筑的侧面设置了一个34.57m的坡道，从梯形广场一端直通二层。使

"Long Residence" is located on the east side of "Xixi Institute" and south to "Bridge Residence", forming an enclosed ladder-shaped outdoor square together with another two buildings named "Bridge Residence" and "Cuboid". It has two storeys with built-up area of 579.5 square meters. As this is the longest building with total length of 66 meters, it is thus named "Long Residence". At the initial stage of design, there was no glass curtain wall on the south side. Originally I intended to reshape the relationship between square and streets of a community in a long and narrow block. In "Xixi Institute", "Plum Flower" is a street and alley community layout in traditional sense, "Trumpet" is extending the street and alley relationship of community further indoors, while "Long Residence" attempts to extend community factors horizontally but be confined by a frame, forming a community layout of street and village pattern. The first storey of "Long Residence" was designed to extend outward, with different staircases leading to small residential rooms of different space on the second storey respectively. These box-like small residential rooms of various types, floating above alleys and streets, constituted richness of single mass. When users climb up and down staircases, they may and can enjoy wetland scenes on the south. However, this conception was somewhat beyond normal daily standard. In order to make the design comply with daily life, we finally decide to add a glass curtain wall outside the semi-open "enframed view city". As a beautiful lake is on the north side of "Long Residence", we set a slope of 34.57 meters in length on its side linking the first and second storey of ladder-shaped square. Users can enjoy ascending or descending wetland scenes

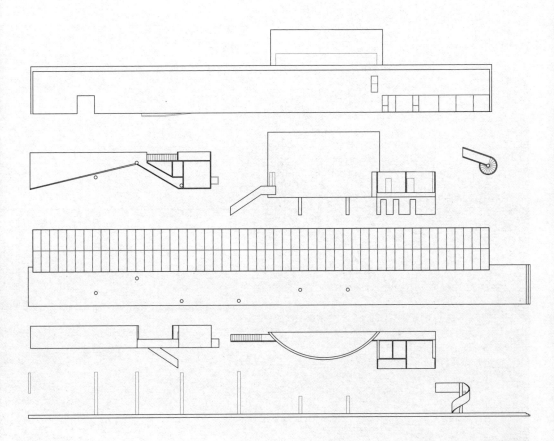

"长宅"的构成要素
Factors constituting "Long Residence"

用者无论是从一层向上抑或是从二层到一层的漫步过程中，都能够感受到湿地风景的逐步升起或降落，进而使得原本平面化的景观在竖向维度上获得一个新的层级与提升。

while they are climbing up or down. Therefore, the originally flat scenes gain a new level and elevation vertically.

"长宅" 北侧风景局部
Partial scenes, north side of the "Long Residence"

"长宅"北侧风景局部
Partial scenes, north side of the "Long Residence"

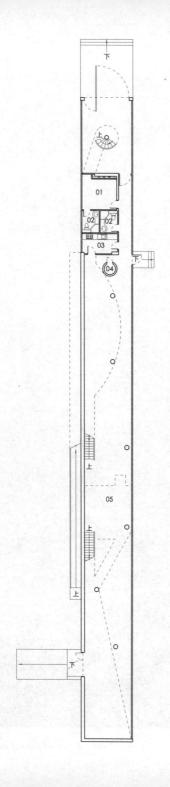

01 卧室　　Bedroom
02 卫生间　Washroom
03 厨房　　Kitchen
04 设备间　Equipment room
05 起居室　Sitting room

一层平面图　1st floor

0　1　　　5m

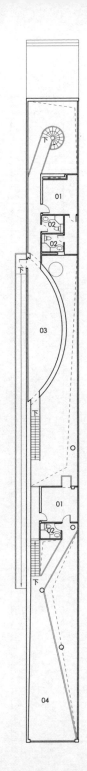

01 卧室　　Bedroom
02 卫生间　Washroom
03 会议室　Conference room
04 起居室　Sitting room

二层平面图　2nd floor

0　1　　5m

N

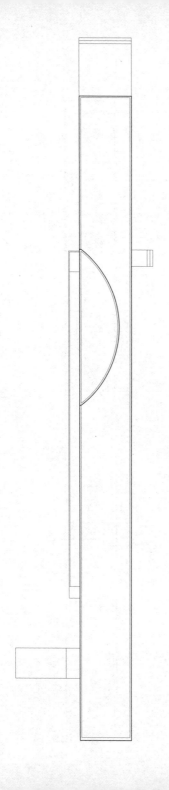

 屋顶平面图 roof plan

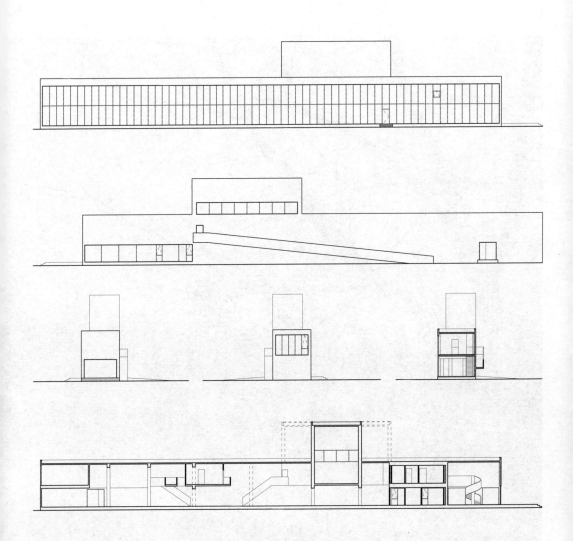

南立面图	south elevation
北立面图	north elevation
东、西立面图、横向剖面图	east&west elevation&cross section
纵向剖面图	longitudinal section

"长宅"南侧风景
South side scenes of "Long Residence"

"长宅" 北广场
"Long Residence"-North Piazza

"长宅"北广场
"Long Residence"-North Piazza

"长宅"北侧风景
North side scenes of "Long Residence"

10. "长方体" "Cuboid"

建筑面积：154.00m²
total floor area：154.00m²

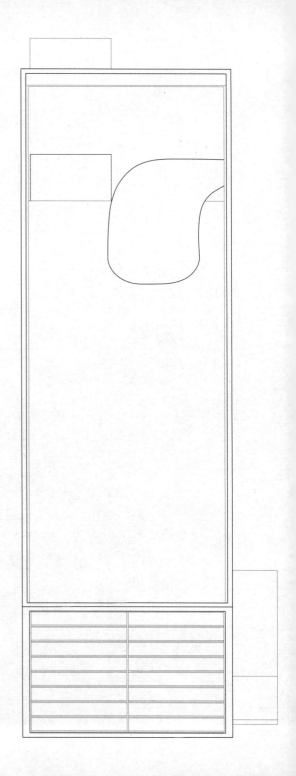

"长方体"轴测图
Axonometric drawing of "Cuboid"

"长方体"建筑,是位于园区最东侧的建筑。建筑长22m,宽7m,高6.76m,整个建筑面积为154m²,是一个多功能报告厅。长方体报告厅建筑的入口在西侧,面向由"桥宅"和"长宅"两个建筑共同围合的梯形广场。

建筑的东侧是一个完全开敞的玻璃幕墙,面对一片优美的湿地景观,目的是让人们坐在这里能够欣赏到湖边的景色。同时,在幕墙的内侧顶部有一个巨大的幕帘,当需要举行活动时,巨大的幕帘就会缓缓地落下,挡住幕墙外部的景色和光线,使得空间的内部形成一个封闭的视觉环境,而幕墙一侧则成为报告厅的舞台。

报告厅内部的座椅是非固定的,这样做的目的是考虑到报告厅建筑的灵活使用问题。这个建筑本身还可以成为一个艺术工作室,或在特别需要的时候,可进行局部二层的加建,建筑的入口是一个4m见方的大门,靠近入口的室内一侧有一个自由造型的装置,装置内部是一个洞窟状的空间场景。从报告厅室内看,装置本身如同是一个自由曲线状的雕塑,这个雕塑装置内部是一个室内卫生间。雕塑装置扭动的体态最终收束到报告厅的北侧墙面上,收束点是一个拱形圆洞,从这里将室外的光线送入洞窟的同时,也与外界之间获得交流与沟通。这个收束点本身也是卫生间的排气窗。

This "Cuboid" building is located on the most east side, with 22 meters in length, 7 meters in width and 6.76 meters in height. It serves as a multi-functional lecture hall with total built-up area of 154 square meters. The building of cuboid lecture hall, whose entrance is on the west side, faces the ladder-shaped squared enclosed by "Bridge Residence" and "Long Residence".

On the east side there is a fully open glass curtain wall facing beautiful wetland scenes so that people can sit and enjoy lakeside views. Meanwhile, there is a huge curtain inside the top of glass wall, which can slowly drop down to block the outside view and light when events are taking place. Therefore an enclosed visual environment is created inside, with the side on glass curtain wall becoming a stage of the lecture hall.

In consideration of flexible use of this building, chairs inside the lecture hall are not fixed. This building can also serve as an art studio or, when necessary in special cases, add an extra second storey partially. At the entrance there is a door of about 4 square meters. On the indoors side near the entrance there is an installation of free modeling, with cave-like scenes inside. Viewed from inside the lecture hall, the installation itself appears like a free-style curvy statue with indoor washroom inside. The twisting posture of this statue is finally converged to the north wall of the lecture hall. The converging point, which is an arch-shaped hole, brings outside light into the hall to gain exchange and communication with the outdoor environment. This converging point is also where the ventilation window of the washroom is located.

"长方体"东南侧风景
South-east side scenes of "Cuboid"

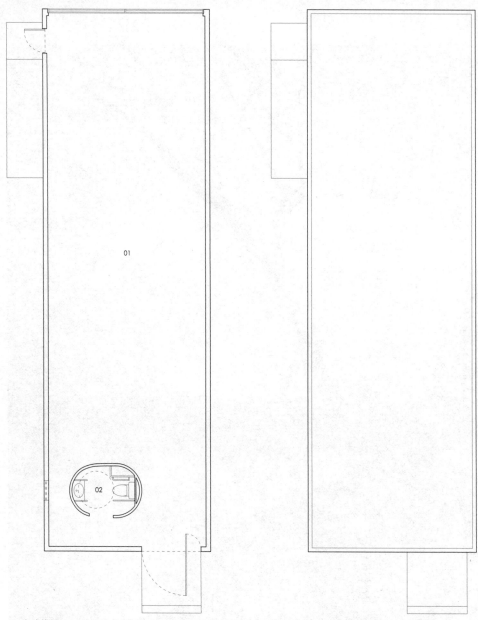

01 多功能厅　Multifunctional hall
02 卫生间　　Washroom

一层平面图　1st floor　　屋顶平面图　roof plan

0　1　　　5m

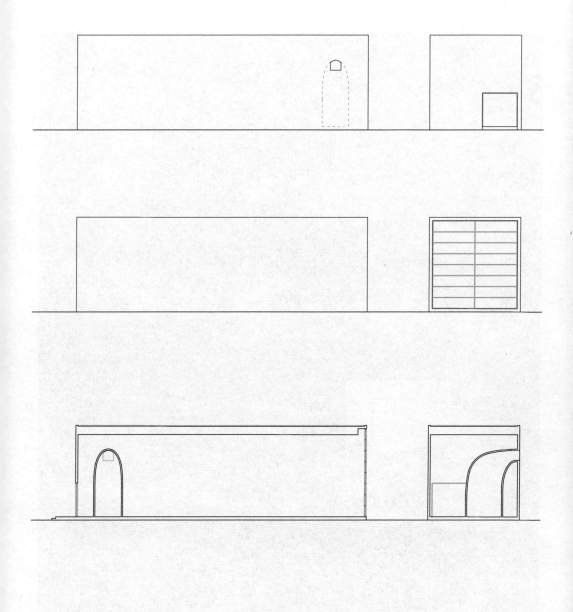

北立面图	north elevation	西立面图	west elevation
南立面图	south elevation	东立面图	east elevation
竖向剖面图	longitudinal section	横向剖面图	cross section

"长方体"室内
Indoors of "Cuboid"

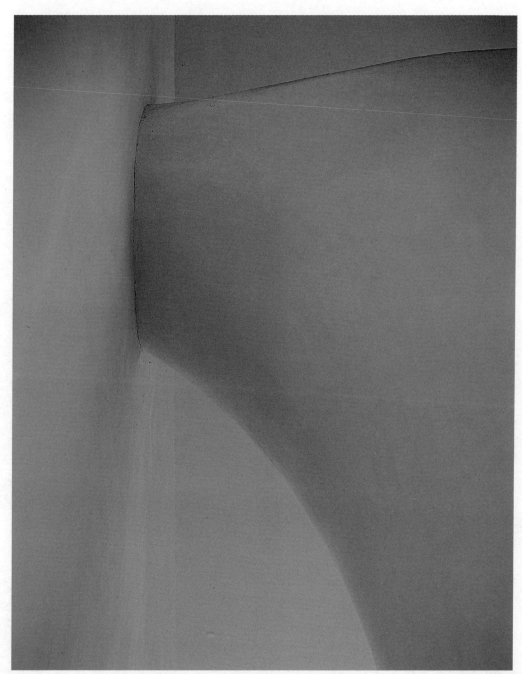

"长方体"室内局部
Interior layout of the "Cuboid"

"长方体"建筑东侧风景
East side scenes of "Cuboid"

"长方体"建筑与周围环境
"Cuboid" and surroundings

11. "水上漂" "Water Float"

建筑面积：217.78m²
total floor area：217.78m²

我以为：在中国传统园林中，水榭的景象表达着一种水上漂的意境。如同杭州西湖中三潭印月整体表达着一种漂浮在水中并造成可移动的动感意境的感受一样。"西溪学社"的"水上漂"建筑便是这样一个意境的表达。

"水上漂"建筑位于基地的北侧湖畔，由10m见方的体块所构成。原本设想让其彻底漂浮在水上，并设想：若要接近它未来只能靠乘船才能到达。然而由于现实的建筑法规上的制约，这个项目本身最终修改成为一个水榭式的做法，即：将建筑靠近北侧的堤岸，主要的入口就从这个建筑的北侧进入，建筑南侧只拥有一个可供船停泊的码头。

"水上漂"建筑的一层是一个大的起居室，沿东西两个景观好的方向开有落地窗。二层是厨房的屋顶，并由此形成一个小的空间转换平台，一层到三层的楼梯从这个平台转换并通往三层。三层整体由三个居室所构成。在三层，设有一个通道，在通道的尽端设有一个旋转楼梯通往屋顶，这个旋转楼梯飘浮在空中，从一层的公共空间向上望去，充满神秘特征。

I reckon that the view of waterside pavilions in Chinese traditional gardens represent an artistic conception of water float. Just like Three Pools Mirroring the Moon as a whole scene in Hangzhou West Lake expresses a kind of floating and moving artistic conception, the building of "Water Float" in "Xixi Institute" also expresses such a conception.
"Water Float", a block of building with about 10 square meters, is located on lakeside on the north side of the base. The original design was to make it totally floating on water and only accessible by boat. However, due to restriction of actual architectural laws and regulations, this is finally revised into a waterside pavilion. To be specific, the building is close to the north bank and accessible through the entrance on the north side of the building, with only a dock for boat to berth on the south.
The first storey of "Water Float" is a big living room, lined with French windows along the east and west side with good view. The second storey is kitchen roof, forming a small spatial conversion platform, through which one can go to the third storey from the first storey by a staircase. The whole third storey consists of three residential rooms and a passageway connected with a spiral staircase leading to roof. This staircase, floating in the air, is full of mysterious feature when one looks up from the public area on the first storey.

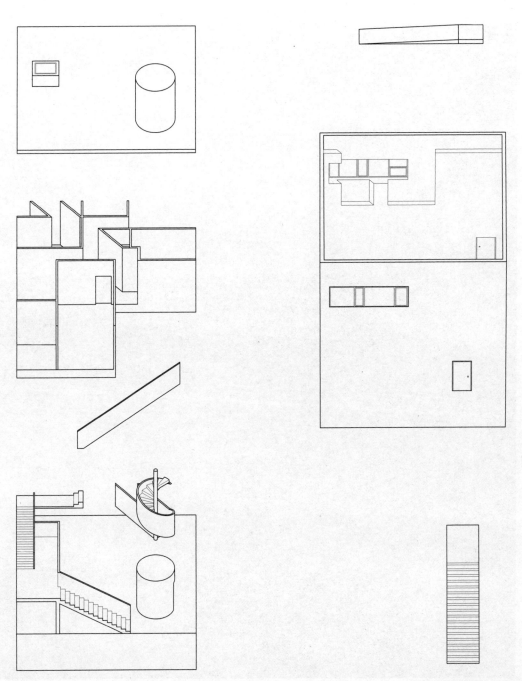

"水上漂"建筑的构成要素
Factors constituting "Water Float"

"水上漂"建筑的南侧风景
South side of "Water Float"

"水上漂"建筑的南立面
South elevation of "Water Float"

"水上漂"建筑的南立面
South elevation of "Water Float"

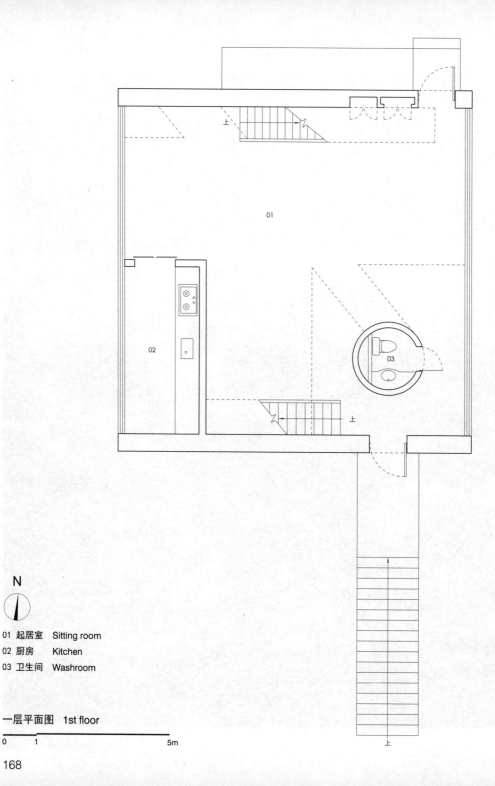

N

01 起居室 Sitting room
02 厨房 Kitchen
03 卫生间 Washroom

一层平面图 1st floor

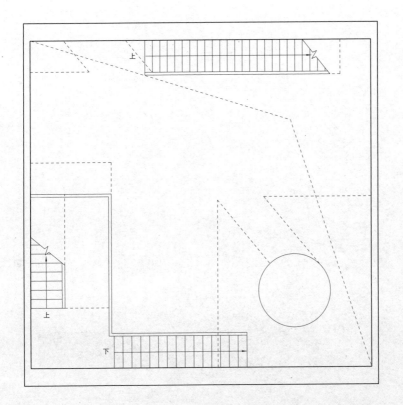

夹层平面图 interlayer floor

"水上漂"建筑的西侧风景
West side scenes of "Water Float"

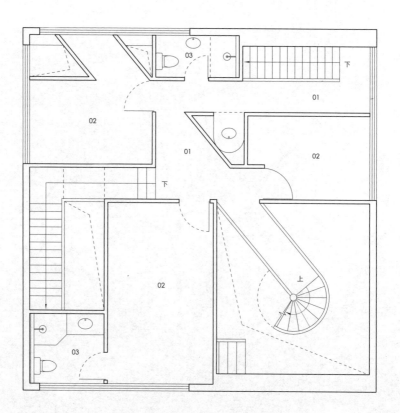

N

01 走廊　Passageway
02 卧室　Bedroom
03 卫生间　Washroom

二层平面图　2nd floor
0　1　　　5m

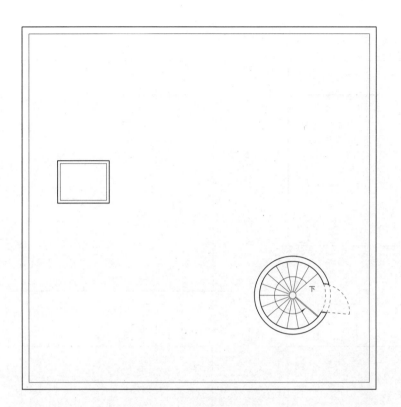

屋顶平面图　roof plan

173

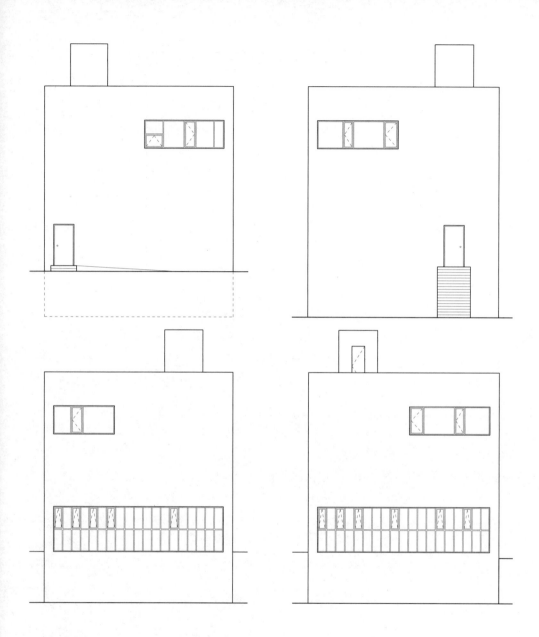

北立面图 north elevation 南立面图 south elevation
西立面图 west elevation 东立面图 east elevation

"水上漂"建筑的西北侧风景
North-west side of "Water Float"

"水上漂" 室内
Indoors of "Water Float"

"水上漂" 室内
Indoors of "Water Float"

"水上漂"室内
Indoors of "Water Float"

通往屋顶的旋转楼梯
Spiral staircase leading to roof

12. "框景舞台" "Enframed Stage"

建筑面积： 21.34m²
total floor area： 21.34m²

南立面图　south elevation
平面图　　plan

0　1　　　　5m

在"椭圆住宅"的北侧，摆放着一个"框"。它框取着湿地北部景观的同时，还是"西溪学社"中的一个"框式"小舞台。舞台9.7m长，2.2m宽，3.4m高。与"椭圆住宅"之间所限定的空场是这个舞台所面向的观众席。同时，二者也可以形成彼此互动的一个舞台布景。

On the north side of the "Oval Residence" a "frame" is placed. It serves not only as a frame for the north wetland scenes but also as a small "frame-style" stage in "Xixi Institute". The size of the stage is 9.7 meters in length, 2.2 meters in width and 3.4 meters in height. Space between the stage and "Oval Residence" is auditorium facing the stage. Meanwhile, these two can also form interactive stage setting.

"框景舞台"
"Enframed Stage"

"框景舞台"与周围建筑
"Enframed Stage" and nearby buildings

"框景舞台"
"Enframed Stage"

"框景舞台"与周围的组合关系
Combination relationship between "Enframed Stage" and surroundings

13. "扇形剧场" "Fan-shaped Theatre"

建筑面积：118.89m²
total floor area：118.89m²

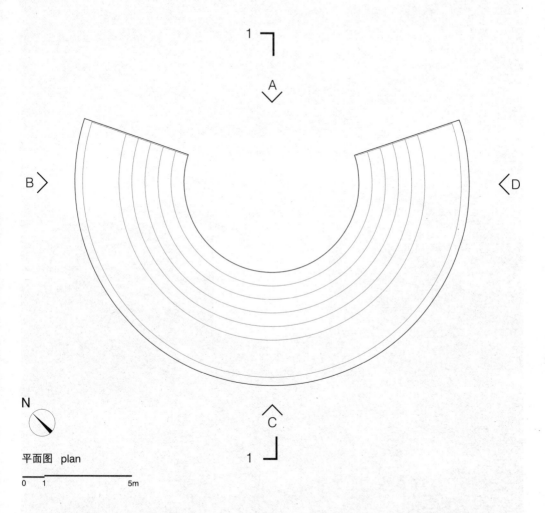

平面图 plan
0 1 5m

"扇形剧场"是"西溪学社"的一个室外小剧场，与位于其东侧的"桥宅"的通向屋顶的大台阶之间共同形成相互呼应的室外剧场。这个室外剧场呈正圆扇面形，扇面的开口部分面向东侧，与由"桥宅"、"长宅"、"长方体"三个建筑所共同围合的广场相呼应

As a small outdoor theatre of "Xixi Institute", this "Fan-shaped Theatre" partners with the big staircase of "Bridge Residence" leading to its roof on its east side to jointly form an outdoor theatre. This outdoor theatre is in the shape of fan from a perfect circle, the opening of the fan faces eastward, not only corresponding to the square enclosed by three buildings, namely "Bridge Residence", "Long Residence "and "Cuboid" but also jointly forming a

"扇形剧场"的西侧风景
West side scenes of "Fan-shaped Theatre"

"扇形剧场"
"Fan-shaped Theatre"

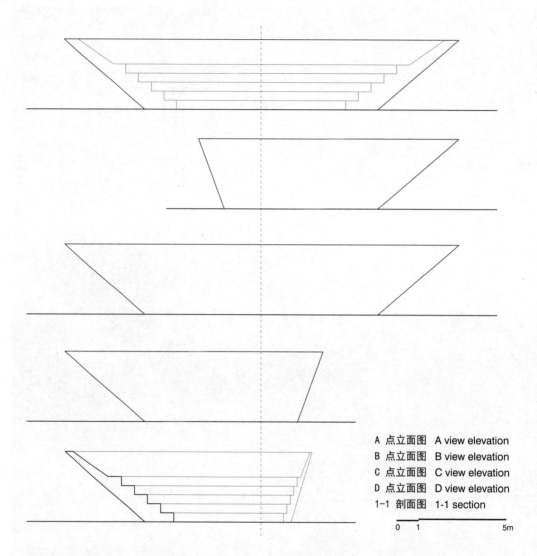

A 点立面图　A view elevation
B 点立面图　B view elevation
C 点立面图　C view elevation
D 点立面图　D view elevation
1-1 剖面图　1-1 section

的同时，还形成一个公共的区域组团。使用者可以集合在这里进行交流和演出，同时也为游人提供一个良好的室外休息场所。小剧场采用的是一个"摆放"的布局方式，将正圆剧场切掉一个角落，打破了剧场本身的封闭性并将剧场开放，这种开放性也使得剧场与"桥宅"之间形成了一个良好的互动和组合。

public regional cluster. Users can gather here for exchange and performance activities. Besides, this is also a good outdoor resting place for visitors. By adopting a layout pattern to cut off one sector from perfect circle theatre, the small theatre is open to the outside instead of self-closure, which forms a good interaction and combination between the theatre and "Bridge Residence".

"扇形剧场"与周围建筑的组合
Combination of "Fan-shaped Theatre" and nearby buildings

"扇形剧场" 西北角
North-west corner of "Fan-shaped Theatre"

作者简介

王昀 博士

1985年	毕业于北京建筑工程学院建筑系获学士学位
1995年	毕业于日本东京大学获工学硕士学位
1999年	毕业于日本东京大学获工学博士学位
2001年	执教于北京大学
2002年	成立方体空间工作室
2013年	创立北京建筑大学建筑设计艺术研究中心担任主任

建筑设计竞赛获奖经历：
1993年日本《新建筑》第20回日新工业建筑设计竞赛获二等奖
1994年日本《新建筑》第4回S×L建筑设计竞赛获一等奖

主要建筑作品：
善美办公楼门厅增建，60m²极小城市，石景山财政局培训中心，庐师山庄，百子湾中学，百子湾幼儿园，杭州西溪湿地艺术村H地块会所等。

参加展览：
2004年6月"'状态'中国青年建筑师8人展"
2004首届中国国际建筑艺术双年展
2006年第二届中国国际建筑艺术双年展
2009年比利时布鲁塞尔"'心造'——中国当代建筑前沿展"
2010年威尼斯建筑艺术双年展，
德国卡尔斯鲁厄Chinese Regional Architectural Creation建筑展
2011年捷克布拉格中国当代建筑展，意大利罗马"向东方-中国建筑景观"展，中国深圳·香港城市建筑双城双年展
2012年第十三届威尼斯国际建筑艺术双年展中国馆等

Profile

Dr. Wang Yun

Graduated with a Bachelor's degree from the Department of Architecture at the Beijing Institute of Architectural Engineering in 1985.
Received his Master's degree in Engineering Science from Tokyo University in 1995.
Received a Ph.D. from Tokyo University in 1999.
Taught at Peking University since 2001.
Founded the Atelier Fronti (www.fronti.cn) in 2002.
Established Graduate School of Architecture Design and Art of Beijing University of Civil Engineering and Architecture in 2013, served as dean.

Prize:
Received the second place prize in the "New Architecture" category at Japan's 20th annual International Architectural Design Competition in 1993
Awarded the first prize in the "New Architecture" category at Japan's 4th SxL International Architectural Design Competition in 1994

Prominent works:
ShanMei Office Building Foyer, A Small City of 60 Square Meters, the Shijingshan Bureau of Finance Training Center, Lushi Mountain Villa, Baiziwan Middle School, Baiziwan Kindergarten, and Block H of the Hangzhou Xixi Wetland Art Village.

Exhibitions:
The 2004 Chinese National Young Architects 8 Man Exhibition, the First China International Architecture Biennale, the Second China International Architecture Biennale in 2006, the "Heart-Made: Cutting-Edge of Chinese Contemporary Architecture" exhibit in Brussels in 2009, the 2010 Architectural Venice Biennale, the Karlsruhe Chinese Regional Architectural Creation exhibition in Germany, the Chinese Contemporary Architecture Exhibition in Prague in 2011, the "Towards the East: Chinese Landscape Architecture" exhibition in Rome, the Hong Kong-Shenzhen Twin Cities Urban Planning Biennale, Pavilion of China The 13th international Architecture Exhibition la Biennale di Venezia in 2012.

图书在版编目（CIP）数据

空间的聚散 / 王昀 著. — 北京：中国建筑工业出版社，
2014.11
ISBN 978-7-112-17434-8

Ⅰ.①空… Ⅱ.①王… Ⅲ.①建筑设计-教材 Ⅳ.①TU2

中国版本图书馆CIP数据核字（2014）第256364号

感谢北京建筑大学建筑设计艺术研究中心建设项目的支持

中 译 英：王 倩
责任编辑：曹 扬
责任校对：陈晶晶 姜小莲
版式设计：张捍平

空间的聚散
王 昀 著

*

中国建筑工业出版社出版、发行（北京西郊百万庄）
各地新华书店、建筑书店经销
北京顺诚彩色印刷有限公司印刷

*

开本：787×1092毫米 1/16 印张：12¾ 字数：243千字
2015年5月第一版 2015年5月第一次印刷
定价：108.00元
ISBN 978-7-112-17434-8
（26254）

版权所有 翻印必究
如有印装质量问题，可寄本社退换
（邮政编码 100037）